Ultrastructure of Rat Adenohypophysis

CORRELATION WITH FUNCTION

ULTRASTRUCTURE of RAT ADENOHYPOPHYSIS

Correlation with Function

Allen Costoff

Department of Endocrinology
Medical College of Georgia
Augusta, Georgia

ACADEMIC PRESS 1973 New York and London

ACADEMIC PRESS, INC.
111 Fifth Avenue, New York, New York 10003

United Kingdom Edition published by
ACADEMIC PRESS, INC. (LONDON) LTD.
24/28 Oval Road, London NW1

LIBRARY OF CONGRESS CATALOG CARD NUMBER: 70-154406

PRINTED IN THE UNITED STATES OF AMERICA

Contents

Chapter 5 Adrenocorticotropes

Chapter 6 Somatotropes

Chapter 7 Mammotropes

Chapter 8 Chromophobes

Chapter 9 Pars Tuberalis

Chapter 10 Pars Intermedia

Chapter 11 Hormonal Activities and Biochemical Properties of Isolated Granule Fractions

Chapter 12 Synthesis, Storage, and Release of Anterior Pituitary Hormones

Preface

During the last fifteen years much information has accumulated on the physiology and biochemistry of the anterior pituitary gland. Corticotropin, somatotropin, and luteotropin of various species have been purified and studied. Progress is being made in the purification of gonadotropins and thyrotropin. In addition, research on mechanisms of hormone action, on synthesis and release of pituitary hormones, and on actions of various blocking agents on pituitary function is being stressed.

These biochemical and physiological findings would be better understood with a knowledge of cellular structure. This book not only deals with the ultrastructure of the anterior pituitary in the normal intact rat but also after various treatments and changes in the physiology of the animal. This is the first time that the complete ultrastructure of the rat pituitary is discussed in a single volume. A book of this kind is needed since the rat is usually one of the principal animals used in research. Wherever possible an attempt has been made to coordinate and associate changes in pituitary ultrastructure with physiological aspects. In order that ultrastructural data be placed in proper perspective, literature on the light microscopy of the rat pituitary gland is reviewed and correlated with ultrastructural findings.

In addition to ultrastructural data, a method is given for the isolation of various populations of pituitary secretory granules that can be used in experimental endocrinological research. The knowledge obtained from this type of granule fractionation and from assays of different fractions of granules has aided in the identification of cell types in the rat interior pituitary gland.

There is a chapter on each of the cell types, including the chromophobe. The often neglected pars tuberalis and pars intermedia of the adenohypophysis are discussed and electron micrographs analyzed. Chapters are also included on the chemical and physical properties of the secretory granules and their formation and release.

This book should serve as a guide for pituitary ultrastructural studies in other species and should lead to a greater understanding of physiological findings. It should be of value not only to the student and the experienced investigator in endocrinology, anatomy, and pathology but also in physiology, biochemistry, and related disciplines.

Much of the experimental work reported here was done in the laboratories of Professors W. H. McShan and R. K. Meyer. I wish to thank them for their encouragement and for reading the manuscript, and Ann Hyatt for the excellent typing job.

Some of the material on granule fractionation and the assay of isolated granule fractions was previously published in the *Journal of Cell Biology* **43** [3], 1969. We express our thanks to the editors for allowing us to reproduce part of this material. I would also like to thank the staff of Academic Press for their valuable help in the production of this book.

This investigation was supported by the National Science Foundation Research Grant No. GB-4963, Public Health Service Training Grant No. 5T1 HD 104-04, Ford Foundation Grant No. 63-505, and the Endocrinology-Reproductive Physiology Training Program, University of Wisconsin.

Allen Costoff

Abbreviations and Terminology Used in the Text

acidophils	cells staining with acid stains (light microscopy) somatotropes, mammotropes, corticotropes (electron microscopy)
amphophils	cells staining with acid and basic stains
AF	aldehyde fuchsin
basophils	cells staining with basic stains (light microscopy) thyrotropes and gonadotropes (electron microscopy)
chromophils	cells containing granules
chromophobes	cells lacking granules
colloid material	nondescript material that stains with lead citrate and uranyl acetate
corticotrope	cell that produces ACTH
dense body	most likely a lysosome
FSH	follicle-stimulating hormone
gonadotropes	cells producing FSH and LH
IU	International Unit
LH	luteinizing hormone
LTH	luteotropin, mammotropin, prolactin, lactogenic hormone
luteotrope	cell producing LTH
M	molar
mammotrope	cell producing LTH
microsomes	broken-up endoplasmic reticulum
mg eq FT	milligram equivalent fresh tissue
m*M*	millimolar
mm	millimeter
mμ	millimicron

MSH	melanocyte-stimulating hormone
mu	milliunit
μ	micron
μC	microcuries
μg	microgram
PAS	periodic acid–Schiff reagent
PTU	propylthiouracil
RF	releasing factors
mRNA	messenger RNA
tRNA	transfer RNA
somatotrope	cell that produces growth hormone
STH	somatotropin, growth hormone
tetrachrome stain	PAS, orange G, alcian blue, alizarin blue
thyrotrope	cell that produces TSH
TSH	thyrotropin
type I cell	most prominant cell type of pars intermedia, 350 mμ granules
type II cell	less abundant cell type of the intermedia, smaller granules

Abbreviations Used in Electron Micrograph Descriptions

CI	cilia
CO	colloid
D	desmosome
DB	dense body (lysosome)
ER	endoplasmic reticulum
F	follicle
G	Golgi complex
LB	light staining body, large bodies
LD	Lipid droplet
M	mitochondria
MB	microbody
MT	microtubule
MV	multivesicular body
MV′	microvilli
P	polysomes
R	ribosomes
RER	rough endoplasmic reticulum
S	secretory granules
V	vacuolar endoplasmic reticulum
VA	vacuoles
VE	vesicles

Chapter 1

Introduction

The histology of the pars distalis of the pituitary gland of rats as well as other species has been studied under normal and various experimental conditions. The object of this work was to relate the different anterior pituitary hormones to a specific cell type in the pars distalis.

Schoenemann (1892) was the first to observe that the secretory granules of the anterior pituitary cells either had a staining affinity for acid or basic dyes or did not stain. The cells were designated as chromophils and chromophobes and the chromophils were divided into acidophils and basophils. A more definitive classification was made possible by application of the periodic acid–Schiff (PAS) reaction which stains gonadotropes and thyrotropes and aldehyde-fuchsin (AF) which also stains thyrotropes. Acidophils have been separated into two cell types by using orange G and azocarmine. Also, autoradiography and immunofluorescence have helped elucidate some cell types at the light microscopic level. A summary of the current views on histochemistry of the pituitary gland is given in Table 1.I.

Reviews of the many staining techniques used in determining the histology of the anterior pituitary gland have been presented by Landing (1954) and by Herlant (1964). The experimental evidence and its interpretation as to cell types has been reviewed by Severinghaus (1937, 1938, 1939), Pearse (1952), Desclin (1952), Hartley (1959), Ezrin (1963), McShan (1965), Holmes (1964), and Vanha-Perttula (1966). Recent general reviews include those by Kracht (1957), Purves (1961, 1966), Herlant (1964, 1965), McShan and Hartley (1965), and Costoff (1969).

TABLE 1.I

A SUMMARY OF CURRENT VIEWS OF RAT ANTERIOR PITUITARY CELL TYPES[a] AND THEIR PROPERTIES

Hormone	Cell type	Staining characteristics[b]				Method of Characterization[b]
		AF	PAS	ACID	Tetrachrome	
Somatotropin	Acidophil Somatotrope	–	–	+ Orange G	Orange G	Acromegaly Tumors Dwarf mice Area of pituitary STH releasers
Luteotropin	Acidophil Luteotrope	–	–	+ Carmine	Brick red	Lactation Tumors Autograft to kidney
Corticotropin	Acidophil Corticotrope	–	–	–	Violet	Disease Blocking Agents Adrenalectomy Cortisone
Thyrotropin	Basophil Thyrotrope	+ Red	+	–	Dark blue	Tumors Thyroxin Thyroidectomy Thiourea
Follicle-stimulating hormone	Basophil Peripheral gonadotrope	–	+ Purple	–	Light blue	Castration Estrogen Estrous cycle Blocking agents
Luteinizing hormone	Basophil Central gonadotrope	–	+ Red	–	Light blue	Castration Estrogen Estrous cycle Blocking agents
No specific hormone	Acidophilic Basophilic chromophobes	–	–	–	–	Stem cells degranulated cells Follicle cells Cleft cells

[a] Many of these cells have been identified using fluorescent-antibody methods.

[b] AF, aldehyde fuchsin; PAS, periodic acid-Schiff; ACID, acid stains; Orange G, azocarmine. Tetrachrome consists of PAS, Orange G, Alcian blue, alizarine blue.

Although much knowledge of the cell types has come from the study of stained sections, the development of electron microscopic techniques is proving of equal or greater value in the study of pituitary morphology. Moreover, by means of electron microscopy the action of certain enzymes can be studied and located at the intracellular level (electron microscopic histochemistry). Other methods have made use of isotopes (high resolution autoradiography), intracellular antibodies coupled with peroxidase (immunoelectron microscopy), and the viewing of sections of cellular fragments after negative staining. Electron microscopic studies of rat pituitaries have been made by Farquhar and Rinehart (1954), Kurosumi and Oota (1968), Fawcett *et al.* (1969), and Siperstein and Miller (1970); of the mouse pituitary, by Barnes (1963); salamander, Cardell (1963); rabbit, Salazar (1963); hamster, Girod and Dubois (1965); and cow, Heath (1970).

Extensive cytological and histological research has been undertaken on anterior pituitaries of many species. This histological work, correlated with physiological experiments, forms the basis for the generally accepted concept that specific cell types synthesize, store, and release each of the anterior pituitary hormones, and that the hormones are stored in or associated with granules. There is general agreement that six hormones are produced by the pars distalis of the adenohypophysis. These hormones are somatotropin (growth hormone, STH); luteotropin (mammotropin, prolactin, LTH); thyrotropin (thyroid stimulating hormone, TSH); follicle-stimulating hormone (FSH); luteinizing hormone (LH) in the female or interstitial cell-stimulating hormone (ICSH) in the male; and corticotropin (adrenocorticotropic hormone, ACTH). The pars intermedia produces melanocyte-stimulating hormone (MSH) and perhaps ACTH. Lipotropic hormone has also been associated with the adenohypophysis in some species.

Cytoplasmic particulate fractions of the pituitary glands from several species of animals have been prepared during recent years, beginning in 1948 with the work of Catchpole (1948). The earliest isolation procedures for the rat were made by McShan and Meyer (1949). Their methods involved differential centrifugation. More recently continuous and discontinuous gradient centrifugation, microfiltration, and chromatographic procedures have been employed to obtain more homogeneous fractions of granules. Electron microscopy is now routinely employed to determine the degree of homogeneity of isolated organelles. Granule fractionation studies coupled with biological assays have proven valuable in identifying the cell types in the rat pituitary, and for determination of the biological, chemical, and physical characteristics of the granules.

In the ensuing chapters, histochemical findings will be reviewed and electron micrographs of pituitary glands from intact and treated rats will be analyzed. The results of granule isolation studies will be presented and coordinated with physiological data. There are also bibliographies for each of the individual chapters.

REFERENCES

Barnes, B. G. (1963). The fine structure of the mouse adenohypophysis in various physiological states. *In* "Cytologie de l'Adenohypophyse" (J. Benoit and C. Da Lage, eds.) , p. 91. Editions du centre National de La Recherche Scientifique, Paris.

Cardell, R. R. (1963). The cytophysiology of the anterior pituitary gland. *Henry Ford Hosp. Med. Bull.* **11**, 409.

Catchpole, H. R. (1948). Cell fractionation and gonadotrophin assays of anterior pituitary glands. *Fed. Proc., Fed. Amer. Soc. Exp. Biol.* **7**, 19.

Costoff, A. (1969). Isolation of secretory granules and study of the fine structure of male rat anterior pituitary glands. Ph.D. Thesis, University of Wisconsin, Madison.

Desclin, L. (1952). The physiological meaning of the histological picture of the anterior hypophysis characteristic of castration, pregnancy, and thyroidectomy. *Ciba Found. Colloq. Endocrinol.* **4**, 21.

Ezrin, C. (1963). "The Pituitary Gland." Ciba Pharmaceutical Co., Summit, New Jersey.

Farquhar, M. G., and Rinehart, J. F. (1954). Cytological alterations in the anterior pituitary gland following thyroidectomy: An electron microscope study. *Endocrinology* **55**, 857.

Hymer, W. C., and McShan, W. H. (1963). Isolation of rat pituitary granules and study

Fawcett, D. W., Long, J. A., and Jones, A. L. (1969). The ultrastructure of endocrine glands. *Recent Progr. Horm. Res.* **25**, 315.

Girod, C., and Dubos, P. (1965). Etude ultrastructurale des cellules gonadotropes antéhypophysaires chez le hamster. *J. Ultrastruct. Res.* **13**, 212.

Hartley, M. W. (1959). Purification, isolation, and description of cytoplasmic granules and particulates associated with gonadotropic hormones of the anterior pituitary gland of the rat. Ph.D. Thesis, University of Wisconsin, Madison.

Heath, E. (1970) . Cytology of the pars anterior of the bovine adenohypophysis. *Amer. J. Anat.* **127**, 131.

Herlant, M. (1964). The cells of the adenohypophysis and their functional significance. *Int. Rev. Cytol.* **17**, 299.

Herlant, M. (1965). Present state of knowledge concerning the cytology of the anterior lobe of the hypophysis. *Proc. Int. Congr. Endocrinol. 2nd, 1964* p. 468.

Holmes, R. L. (1964). Experimental, histochemical and ultrastructural contributions to our understanding of mammalian pituitary function. *Int. Rev. Gen. Exp. Zool.* **1**, 187.

Kracht, J. (1957). Zur lokalisation der hypophysenvorderlappenhormone. *Zentralbl. Allg. Pathol. Pathol. Anat.* **97**, 24.

Kurosumi, K., and Oota, Y. (1968). Electron microscopy of two types of gonadotrophs in the anterior pituitary glands of persistent estrous and diestrous rats. *Z. Zellforsch. Mikrosk, Anat.* **85**, 34.

Landing, B. H. (1954). Histologic study of the anterior pituitary gland; Compilation of procedures. *Lab. Invest.* **3,** 348.

McShan, W. H. (1965). Ultrastructure and function of the anterior pituitary gland. *Proc. Int. Congr. Endocrinol., 2nd, 1964* p. 382.

McShan, W. H., and Hartley, M. W. (1965) Production, storage and release of anterior pituitary hormones *Ergeb. Physiol., Biol. Chem. Exp. Pharmakol.* **56,** 264.

McShan, W. H., and Meyer, R. K. (1949). Gonadotropic activity of granules isolated from rat pituitary glands. *Proc. Soc. Exp. Biol. Med.* **71,** 407.

Pearse, A. G. E. (1952). Cytochemical localization of the protein hormones of the anterior hypophysis. *Ciba Found. Colloq. Endocrinol.* **4,** 1.

Purves, H. D. (1961). Morphology of the hypophysis related to its function. *In* "Sex and Internal Secretions" (W. Young, ed.), 3rd ed., Vol. 1, p. 161. Williams & Wilkins, Baltimore, Maryland.

Purves, H. D. (1966). *In* "The Pituitary" (W. Harris and B. Donovan, eds.), Vol. 1, p. 147. Butterworth, London.

Salazar, H. (1963). The pars distalis of the female rabbit hypophysis, an electron microscopic study. *Anat. Rec.* **147,** 469.

Schoennemann, A. (1892). Hypophysis and thyreoidea. *Arch. Pathol. Anat. Physiol. Klin. Med.* **129,** 310.

Severinghaus, A. E. (1937). Cellular changes in the anterior hypophysis with special reference to its secretory activities. *Physiol. Rev.* **17,** 556.

Severinghaus, A. E. (1938). Cytology of the pituitary glands. *Res. Publ., Ass. Res. Nerv. Ment. Dis.* **17,** 69.

Severinghaus, A. E. (1939). Anterior hypophyseal cytology in relation to the reproductive hormones. *In* "Sex and Internal Secretions" (E. Allen, ed.), 2nd ed., p. 1045. Williams & Wilkins, Baltimore, Maryland.

Siperstein, E. R., and Miller, K. J. (1970). Further cytophysiologic evidence for the identity of the cells that produce adrenocorticotrophic hormone. *Endocrinology* **86,** 451.

Vanha-Perttula, T. (1966). Esterases of the rat adenohypophysis. *Acta Physiol. Scand.* **69,** Suppl. 283. 104 pp.

Chapter 2

Experimental Procedures

ANIMALS

Holtzman strain male and female rats weighing 300–350 gm and 90 days old were used in the experiments to be described. Male rats were killed 1, 2, 3, 4, 5, 7, 14, 30, and 60 days after adrenalectomy and 7, 14, 30, and 60 days after castration. These rats were kept on a normal rat regimen supplemented with 0.9% sodium chloride and 5% sucrose added to the drinking water.

Rats which were both adrenalectomized and castrated were killed 7, 14, and 30 days after treatment. Rats were treated with the compounds in the following amounts and volumes per injection: *(a)* propylthiouracil, 5 mg/0.5 ml; *(b)* Metopirone, 7.5 mg/0.5 ml; *(c)* amphenone [1,1-dichloro-2,2-bis-(*p*-chlorophenyl) ethane], 10 mg/0.5 ml; *(d)* α-ethyltryptamine, 3 mg/0.3 ml; *(e)* ACTH, 0.5 mg/0.5 ml; and *(f)* dexamethasone, 1.0 mg/0.3 ml.

These compounds were either dissolved or suspended in 0.25% methyl cellulose, with the exception of dexamethasone which was dissolved in corn oil. Three injections of 0.5 ml each were given subcutaneously daily over a 3-day period, and the animals were autopsied 2 hours after the last injection.

A group of male rats was given 1.0 mg of hydrocortisone in 0.2 ml of corn oil per day. These animals were injected with hydrocortisone for 1, 2, 7, 14, 30, and 60 days, and were autopsied 24 hours after the final injection.

Female and male rats were hypophysectomized and pieces of pituitary were autografted under the kidney capsule. Grafts were removed 14 days later.

The pituitary glands of female rats were removed during proestrus, estrus, metestrus, diestrus, and on the sixteenth day of pregnancy.

PREPARATION OF TISSUES FOR ELECTRON MICROSCOPIC STUDY

The fixatives used were Zetterquist (1956) ion implemented veronal acetate, or Karnovsky (1965) phosphate buffered glutaraldehyde as revised by Ris (1966). The embedding medium was the Epon-Araldite formula of Mollenhauer (1964) as revised by Ris (1966). The final polymerization was at 38° C for 72–96 hours.

Thin sections were prepared on the Porter-Blum MT–2 ultramicrotome and placed on Formvar coated grids that were usually carboned. Sections of whole tissue were made from different parts of the pituitary gland (Fig. 2.1). With an alternate thick–thin sectioning technique, the greater part of a pituitary was scanned. At least fifty grids containing about one hundred sections each were taken from each part of the pituitary gland.

The sections were stained with uranyl acetate and counterstained with lead citrate (Reynolds, 1963). Sections were studied by use of a Hitachi HS-7S electron microscope equipped with a high voltage stabilizer and double condenser. An accelerating potential of 50 KV and a 50μ objective aperture were used. Magnification was calibrated with a carbon replica of a diffraction grating (54, 864 lines per inch, Fullam Co., Schenectady, New York).

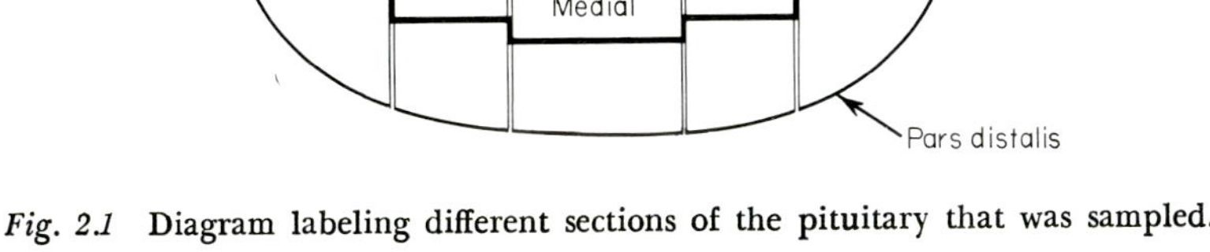

Fig. 2.1 Diagram labeling different sections of the pituitary that was sampled.

PREPARATION OF PITUITARY GRANULES

The pituitary glands were obtained from rats similar to those used for electron microscopy. The animals were killed by cervical dislocation; the pituitary glands were removed immediately, and the anterior lobes were separated from the posterior and placed in a cooled humidor. A 5% homogenate was made in a medium consisting of 0.25 *M* sucrose and 7.3% polyvinylpyrrolidone (PVP) adjusted to pH 7.4. Homogenization was performed in a cold, sharp-pointed glass homogenizing tube. The glands were mashed rather than ground to minimize breaking of granules and solubilizing the hormones.

The whole homogenate was fractionated at 4° C in a multispeed head of a refrigerated Model PR-2 International centrifuge. The nuclear pellet (NP) was obtainted by centrifugation at 275 *g* for 10 minutes. The resulting supernatant (S1) was filtered through a SS Millipore filter with a pore diameter of 3.0μ in a Swinny adapter fastened to a 5-ml Luer-Lok syringe. This filtered supernatant (FS1) was layered on the continuous density gradient.

The gradients were prepared by a modification of the method of Hartley *et al.* (1960). Diodrast, the diethanolamine salt of 3,3-diiodo-4-pyridone-*N*-acetic acid which is available (Winthrop Laboratories, Evanston, Illinois) as a 35% (w/v) aqueous solution, was mixed with sucrose for making the gradients (Otteson and Weber, 1955). The solutions were 6 and 45% (gm/100 ml) sucrose, containing 17.5% (gm/100 ml) Diodrast and 5×10^{-4} *M* Versene. The pH of these solutions was adjusted to 7.2 with potassium hydroxide. The gradients were prepared by use of the apparatus reported by Sinclair *et al.* (1957). In preparing the gradients the apparatus was calibrated to deliver 4.4 ml into each tube. The density of the gradient ranged from 1.11 at the top to 1.26 at the bottom.

One ml of the FS1 fraction was layered onto the surface of the gradient with a 2-ml syringe. The tubes were placed in a SW-39 swinging bucket head and centrifuged in a Beckman Spinco Model L centrifuge for 2 hours at 100,000 *g*. The zones of the gradient were individually removed with a long needle attached to a 2 ml syringe placed in a holder for removal of the desired zone.

The zones were removed starting at the top of the gradient and were identified as shown in Fig. 2.2. The SA layer containing hormones in solution was removed and used for assays. The other zones were diluted with 0.88 *M* sucrose prior to centrifugation for recovery of the particles.

Zones B, C, and D1 were individually removed, diluted and centrifuged at 33,000 rpm (100,000 *g*) for 1 hour to obtain the respective high-speed pellets. Zones D2 and E1 were removed and slowly filtered through moistened Nuclepore filters (General Electric Co., Pleasanton, California) with 0.5μ diameter pores mounted in a Swinny adapter at-

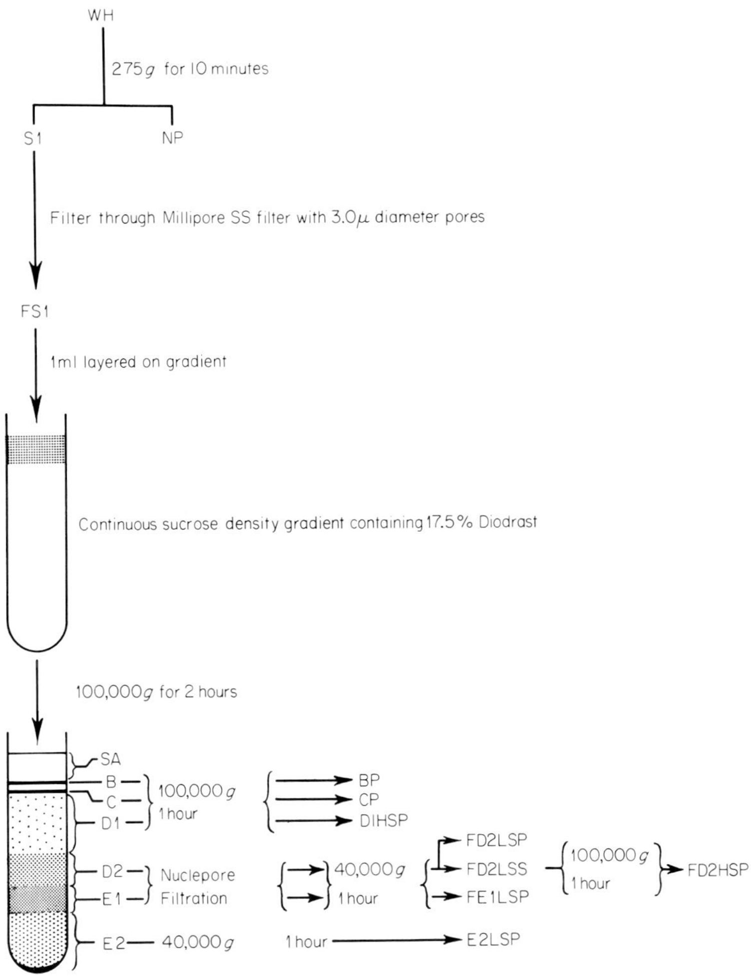

Fig. 2.2. Procedure for isolation of rat pituitary secretory granules. Designation of fractions: WH, whole homogenate (50 mg FT per ml); NP, nuclear pellet, S1, supernatant from nuclear pellet; FS1, filtered supernatant, SA to E2, zones obtained on the continuous density gradient; SA, soluble A zone; BP, ribosomal pellet; CP, microsomal pellet; D1HSP, D1 high-speed pellet; FD2LSP, filtered D2 low-speed pellet; FD2LSS, filtered D2 low-speed supernatant; FD2HSP, filtered D2 high-speed pellet; FE1LSP, filtered E1 low-speed pellet; E2LSP, E2 low-speed pellet.

tached to a 2.0-ml Luer-Lok syringe. These filtered solutions, FD2 and FE1, were diluted and centrifuged at 17,000 rpm (40,000 *g*). The resulting filtered low-speed pellets were designated as FD2LSP and FE1LSP. When the supernatant (FD2LSS) from the FD2LSP pellet was centrifuged for 1 hour at 100,000 *g*, a small pellet FD2HSP was obtained. The lowest zone E2 which usually included a diffuse pellet at the bottom of the tube was removed, diluted, and centrifuged at 40,000 *g* for 1 hour to obtain the pellets which were recovered and used for electron microscopy or were diluted with saline for bioassays.

The pellets obtained during the fractionation were fixed overnight at 4° C in 1% osmic acid containing sucrose (Palade and Siekevitz, 1956). After dehydration the pellets were embedded in methacrylate (Farquhar, 1956) or Epon-Araldite (Mollenhauer, 1964). Each pellet was sectioned at different levels to determine whether stratification had occurred.

The diameters of sections of isolated granules and granules in whole tissue for each of the cell types were measured with a 7× magnifier with 0.1-mm scale divisions. The longest axes of the granules were measured to the outside of the membrane on micrographs usually enlarged ×30,000.

BIOLOGICAL ASSAYS

Zones from the gradient were diluted and centrifuged, and the pellets obtained were suspended in 0.9% saline for bioassay. The assay data were analyzed by the parallel line method of Bliss (1952). Fractions from two to four gradients were required for the assay of each hormone.

Thyrotropin (TSH) was assayed by a modification of the ^{32}P uptake assays of Lamberg (1953) and Greenspan *et al.* (1956). Twenty μc of ^{32}P diluted with saline were administered to 1-day-old chicks.

Adrenocorticotropin (ACTH) was determined by the method of Saffran and Schally (1955).

Follicle-stimulating hormone (FSH) was assayed by the method of Steelman and Pohley (1953).

Luteinizing hormone (LH) was determined by the ovarian ascorbic acid depletion method of Parlow (1961). The determinations of total ascorbic acid was performed according to the modification reported by Schaffert and Kingsley (1955). The slope of the regression of ascorbic acid content on ovarian weight was calculated by analysis of covariance (Steel and Torrie, 1960). Estimates of potency of LH preparations were calculated by the parallel line method of Bliss (1952).

The *somatotropin* (STH) content of the fractions was determined by a modification of the procedure of Greenspan *et al.* (1949). Female rats hypophysectomized at 25 days of age were obtained from the Hormone Assay Laboratories, Chicago, Illinois. Beginning at 30 days of age, the animals were injected with the granule preparations twice daily for 3 days, and autopsies were performed 25 hours after the last injection.

Prolactin (LTH) activity of the granule fractions was measured by a modification (Costoff, 1969) of the Mishkinsky *et al.* method (1967). Rat inguinal mammary glands were cultured in disposable, sterile 35 × 10 mm Petri dishes containing supplemented M-199 medium. The Petri dishes were placed in a closed, moist incubator at 38° C, flushed with 95% O_2–5% CO_2, and removed after 48 hours. A positive alveolar response of the hematoxylin-stained explant is the basis for this assay.

The data obtained from the study of the biological activity of the isolated granules is presented in the chapters concerned with the description of each of the cell types.

The biological and biochemical properties of the isolated granules are discussed in Chapter 11.

REFERENCES

Bliss, C. I. (1952). "The Statistics of Bioassay." Academic Press, New York.

Costoff, A. (1969). Isolation of secretory granules and study of the fine structure of male rat anterior pituitary glands. Ph.D. Thesis, University of Wisconsin, Madison.

Farquhar, M. G. (1956). Review and compilation of procedures for election microscopy. *Lab. Invest.* **5**, 317.

Greenspan, F. S., Li, C. H., Simpson, M. E., and Evans, H. M. (1949). Bioassay of hypophyseal growth hormone: The tibia test. *Endocrinology* **45**, 455.

Greenspan, F. S., Kriss, J. P., Moses, L. E., and Lew, W. (1956). An improved bioassay method for thyrotropic hormone using thyroid uptake of radiophosphorus. *Endocrinology* **58**, 767.

Hartley, M. W., McShan, W. H., and Ris, H. (1960). Isolation of cytoplasmic pituitary granules with gonadotropic activity. *J. Biophys. Biochem. Cytol.* **7**, 209.

Karnovsky, M. J. (1965). A formaldehyde–glutaraldehyde fixative of high osmolality for use in electron microscopy. *J. Cell Biol.* **27**, 137A.

Lamberg, B. (1953). Radioactive phosphorus as indicator in a chick assay of TSH. *Acta Med. Scand.* 145, Suppl. 279. 79 pp.

Mishkinsky, J., Dikstein, S., Ben-David, M., Azeroual, J., and Sulman, F. G. (1967). A sensitive *in vitro* method for prolactin determination. *Proc. Soc. Exp. Biol. Med.* **125**, 360.

Mollenhauer, H. H. (1964). Plastic embedding mixtures for use in electron microscopy. *Stain Technol.* **39**, 111.

Ottesen, M., and Weber, R. (1955). Density-gradient centrifugation as a means of separating cytoplasmic particles. *C. R. Trav. Lab. Carlsberg, Ser. Chim.* **29**, 417.

Palade, G. E., and Siekevitz, P. (1956). Liver microsomes. *J. Biophys. Biochem Cytol.* **2**, 171.

Parlow, A. F. (1961). Bioassay of LH. *In* "Human Pituitary Gonadotropins" (A. Albert, ed.), p. 300. Thomas, Springfield, Illinois.

Reynolds, E. J. (1963). The use of lead citrate at high pH as an electron opaque stain for electron microscopy. *J. Cell Biol.* **17**, 208.

Ris, H. (1966). Personal communication.

Saffran, M., and Schally, A. V. (1955). *In vitro* bioassay of corticotropin: Modification and statistical treatment. *Endocrinology* **56**, 523.

Schaffert, R. R., and Kingsley, G. R. (1955). A rapid simple method for the determination of reduced, dehydro- and total ascorbic acid in biological material. *J. Biol. Chem.* **212**, 59.

Sinclair, J. B., Geil, P. H., and Kaisberg, P. (1957). Biophysical studies of wild cucumber mosaic virus. *Phytopathology* **47**, 372.

Steel, R. G., and Torrie, J. H. (1960). "Principles and Procedures of Statistics," p. 312. McGraw-Hill, New York.

Steelman, S. L., and Pohley, F. M. (1953). Assay of the follicle stimulating hormone based on the augmentation with human chorionic gonadotropin. *Endocrinology* **53**, 604.

Zetterquist, H. (1956). The ultrastructural organization of columnar absorbing cells of the mouse jejunum. Dep. Anat., Ph. D. Thesis, Karolinska Institut, Stockholm.

Chapter 3

Thyrotropes

LIGHT MICROSCOPY

Pituitary thyrotropes were first identified by light microscopy of cellular changes in the pituitary gland that result from hyperthyroidism and thyroidectomy. These changes were described by Hohlweg and Junkman (1933), Severinghaus (1935), Severinghaus *et al.* (1934), and Zechwer *et al.* (1935). The problem arose, however, concerning which cell type produced TSH because degranulation of acidophils and hypertrophy of certain basolphils occurred after thyroidectomy. Severinghaus (1937) concluded that TSH is secreted by acidophils, but Zechwer *et al.* (1935) and Griesbach (1941) reported TSH is associated with basolphils. Guyer and Claus (1937) postulated that TSH basophils were distinct from castration cells. Herlant (1943) used cytochemical methods to study the problem of pituitary cell specificity. Mucoproteins were known to stain with toluidine blue. When Herlant employed this staining method, he found that granules in TSH, FSH, and LH cells stained positively. This indicated that these hormones are mucoproteins as compared with the serous proteins, LTH and STH.

Griesbach and Purves (1945) noted changes in certain acidophils and basophils after thyroidectomy. They determined the normal thyroxine requirement of thyroidectomized rats and maintained them on slightly subnormal levels. Hyperplasia and increased activity occurred in certain basophils while acidophils remained unaltered. It was concluded that TSH is secreted by basophils.

The periodic acid–Schiff (PAS) reaction developed by McManus

(1946) and Hotchkiss (1948) has been used to demonstrate the presence of mucoprotein in cells. Catchpole (1949) and Pearse (1948) were the first to apply the PAS reaction to stain mucoproteins in the hypophysis. Pearse (1949) showed PAS-positive material was present in basophils and certain basophilic chromophobes in the pituitary of all species of animals studied.

Using orange G with trichrome stains Pearse (1950) found that basophilic granules stained red and acidophilic granules, yellow. Purves and Griesbach (1951a,b) used PAS staining to distinguish two different types of basophils. Polyhedral cells in the center of the gland were thought to secrete TSH because they underwent hypertrophy in thyroidectomized animals. Only these cells decreased in size after thyroxine was administered to intact rats. The basophils that did not change under the above conditions were designated as gonadotropes.

The aldehyde fuchsin (AF) stain of Gomori (1950) was first used by Halmi (1950). Purves and Griesbach (1951a,b) stained pituitaries with AF and identified as thyrotropes the cells whose granules stained red. This work has been substantiated by Halmi (1952) and Dhom and Tietze (1962). Purves (1956) reported that thyroidectomy cells were modified thyrotropes.

In 1960 Herlant observed that only TSH granules stain blue with alcian blue dye. With aldehyde-thionin only thyrotropic granules gave a positive reaction (Paget and Eccleston, 1960; Ezrin and Murray, 1963). Pasteels and Herlant (1963) found that TSH granules specifically stained dark blue with the tetrachrome staining method. Deminatti (1962) reported the positive PAS reaction in thyrotropes was due to the presence of sulfate groups in the TSH granules. Pituitary glands were studied *in vitro* and in sections by Fand and Thorell (1962) who used the PAS method in conjunction with spectrophotometry to identify TSH cells. In 1964 Herlant reported that the granules of TSH cells contained more mucoproteinaceous compounds than gonadotropic granules and that this accounted for the granules staining more intensely with PAS.

Greenspan and Hargadine (1965) identified the source of TSH using immunofluorescence techniques. Vanha-Perttula (1966) reported that thyrotropes, AF positive cells, contain fine granules and are often found in groups bordering capillaries. Using peroxidase-labeled antibody methods, Nakane (1968) observed that the cells which reacted with anti-TSH were more numerous in the center of the gland and were often found in clusters.

In summary, thyrotropes have been identified using several histochemical methods. This identification has been confirmed by immunofluorescence studies.

ULTRASTRUCTURE OF THYROTROPES

The thyrotrope was the first pituitary cell type to be definitively identified by electron microscopy. Farquhar and Rinehart (1954) studied rat pituitary glands before and at different times after thyroidectomy and observed that only cells postulated to be the source of TSH hypertrophied. Studies by Kurosumi and Oota (1966) distinguished thyrotropes from gonadotropes and corticotropes. Greatly stimulated TSH cells have been found in naturally occurring and experimentally induced tumors by Farquhar and Furth (1959), Theret and Renault (1964), Feltkamp and Kwa (1965), and Messier (1965).

The Thyrotrope in Untreated Rats

The thyrotrope is perhaps the least abundant of the cell types in the rat pituitary gland as only 1–2% of the cells are of this type. These are also the smallest type of cells, with the possible exception of certain chromophobes. Most of these cells are located in the midcentral and midlateral parts of the gland.

Thyrotropes usually are located on a capillary. They are angular in shape, have little cytoplasm and can be further characterized by the presence of many granules arranged around the periphery of the cell near the cell membrane (Fig. 3.1). The diameters of sections of granules in TSH cells range from 40–150 mu with a mean of 88 mu. Although these granules are less dense than the acidophilic granules, their membranes adhere closely to the granule contents just as they do to acidophilic granules.

This cell lacks extensive fine structure. Centrioles with cilia are sometimes observed near the cell membrane. Although the endoplasmic reticulum is poorly developed, occasionally there are several small cisternae to which ribosomes are intermittently attached; however, the endoplasmic reticulum is usually vesicular in appearance.

The Golgi apparatus is located near the nucleus and, although rather small, it is proportional to the cell volume. Vesicles and vacuoles, usually flattened, and small granules in various stages of formation are found within the Golgi complex. Lysosomes are seldom seen. Mitochondria, ovoid or short rod-shaped bodies containing interrupted cristae randomly oriented are not found in these cells.

The Thyrotrope after Propylthiouracil (PTU) Treatment

After treatment with PTU the rat thyroid retains its ability to accumulate iodide but there is an inhibition of the initial monoiodination of tyrosine. This compound causes inhibition of diiodination of tyrosyl groups

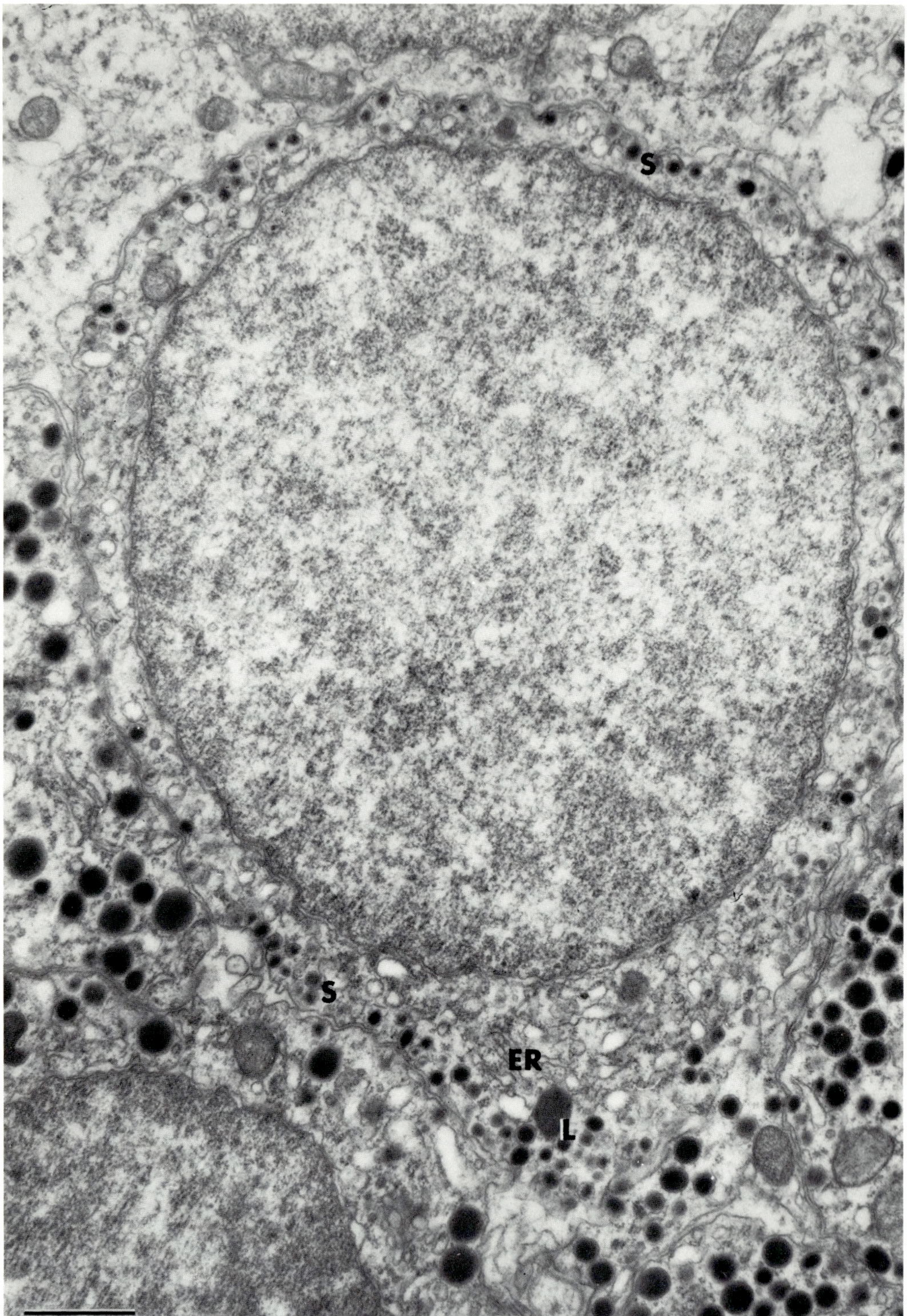

Fig. 3.1. A thyrotrope from the pituitary of a male rat. This small cell has little cytoplasm. Granules (S) of varying electron density are found near the cell membrane. A lysosome (L) and poorly developed vesicular endoplasmic reticulum (ER) are shown. × 17,440. (Veronal acetate buffered osmic acid)

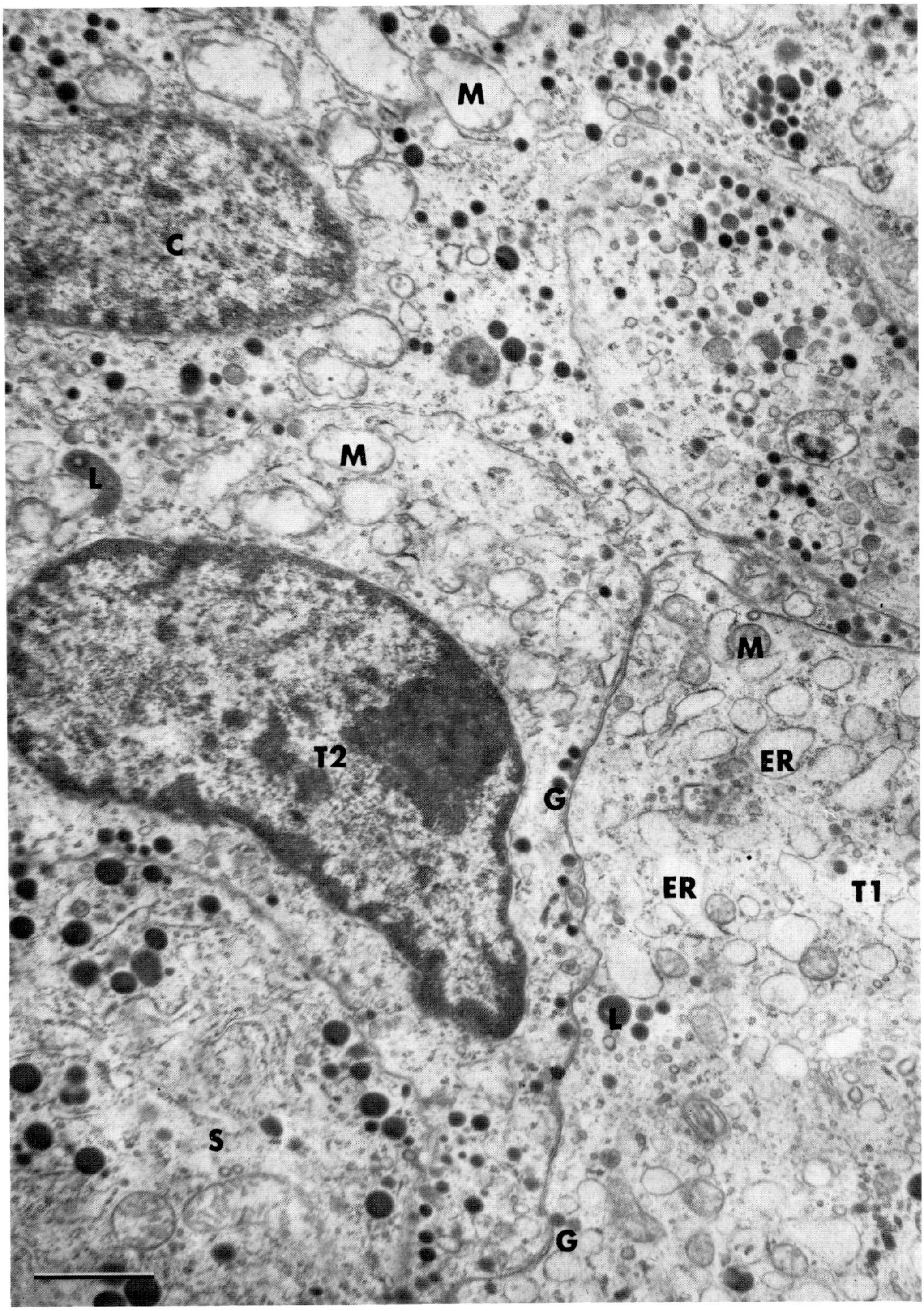

Fig. 3.2. Thyrotropes of a male rat pituitary gland after propylthiouracil (PTU) treatment. Thyrotropes (T1, T2) , a somatotrope (S) and acorticotrope (C) are shown. Thyrotrope (T2) is beginning to hypertrophy after PTU treatment. A lysosome (L), swollen mitochondria (M), and granules (G) are seen in this cell. Thyrotrope (T1) has normal appearing mitochondria (M), lysosomes (L), a vacuolar endoplasmic reticulum (ER), and granules (G) along the cell periphery. The somatotrope (S) appears like those found in control pituitaries. The corticotrope (C) is also similar to those found in the control rat pituitary with the exception of hypertrophied mitochondria (M). × 18,000. (Veronal acetate buffered osmic acid)

and coupling of two tyrosines to form triiodotyrosine and thyroxine (Slingeiland *et al.* 1959; Richards and Ingbar, 1959). Iino *et al.* (1961) showed that daily subcutaneous doses of 5–15 mg PTU were sufficient to prevent the coupling of two tyrosyl groups. As early as 1953 Bogdanove and Halmi observed that daily administration of 10–20 mg of propylthiouracil to rats resulted in an increase of thyroidectomy cells. They concluded that PTU treatment simulates thyroidectomy.

After PTU treatment the angular thyrotropes gradually change size and shape (Fig. 3.2), and they hypertrophy up to tenfold in size (Fig. 3.3). There appears to be an increase in number of TSH cells after this treatment. The hypertrophy of the stimulated TSH cells is due to the great dilation of the rough endoplasmic reticulum. The vesicles fuse, and the result is the characteristic vacuolated endoplasmic reticulum (Fig. 3.4) of the thyroidectomy cell. There is a paucity of granules but the Golgi apparatus is hypertrophied and better defined indicating an actively synthesizing cell which is secreting more hormone than it stores.

The mitochondria of the stimulated TSH cells are generally hypertrophied; oval or spherical types (Fig. 3.4) are more prevalent. Because of the great enlargement of the endoplasmic reticulum, the nuclei are often indented and sometimes an increased number of nucleoli are found in the nuclei. Fewer granules are present in these cells, but their mean diameters are 40 mμ greater than those in untreated TSH cells.

DISCUSSION

Contopoulos and co-workers (1958) also noted fewer granules in TSH cells after treatment with PTU as indicated by light microscopy. D'Angelo (1961) assayed blood and pituitary glands from rats after PTU treatment and found that TSH increased 300% in the blood but was reduced by 95% in the pituitary. These results suggest that there is a negative feedback of thyroxine on the pituitary or hypothalamus which inhibits TSH release and promotes storage of TSH in pituitary granules. However, after thyroidectomy or PTU administration, little thyroxine is produced; TSH is continually secreted into the blood but little is stored.

Barnes (1963) reported dilated endoplasmic reticulum and a decrease in granulation in the hypertrophied thyrotropes after thyroidectomy of mice. When thyroxine was administered, a regranulation of these cells occurred. Potvliege (1970) showed electron micrographs of thyroidectomy cells after PTU treatment of rats. When he administered PTU with estrogen, he also observed a population of "regular intact" TSH cells and a

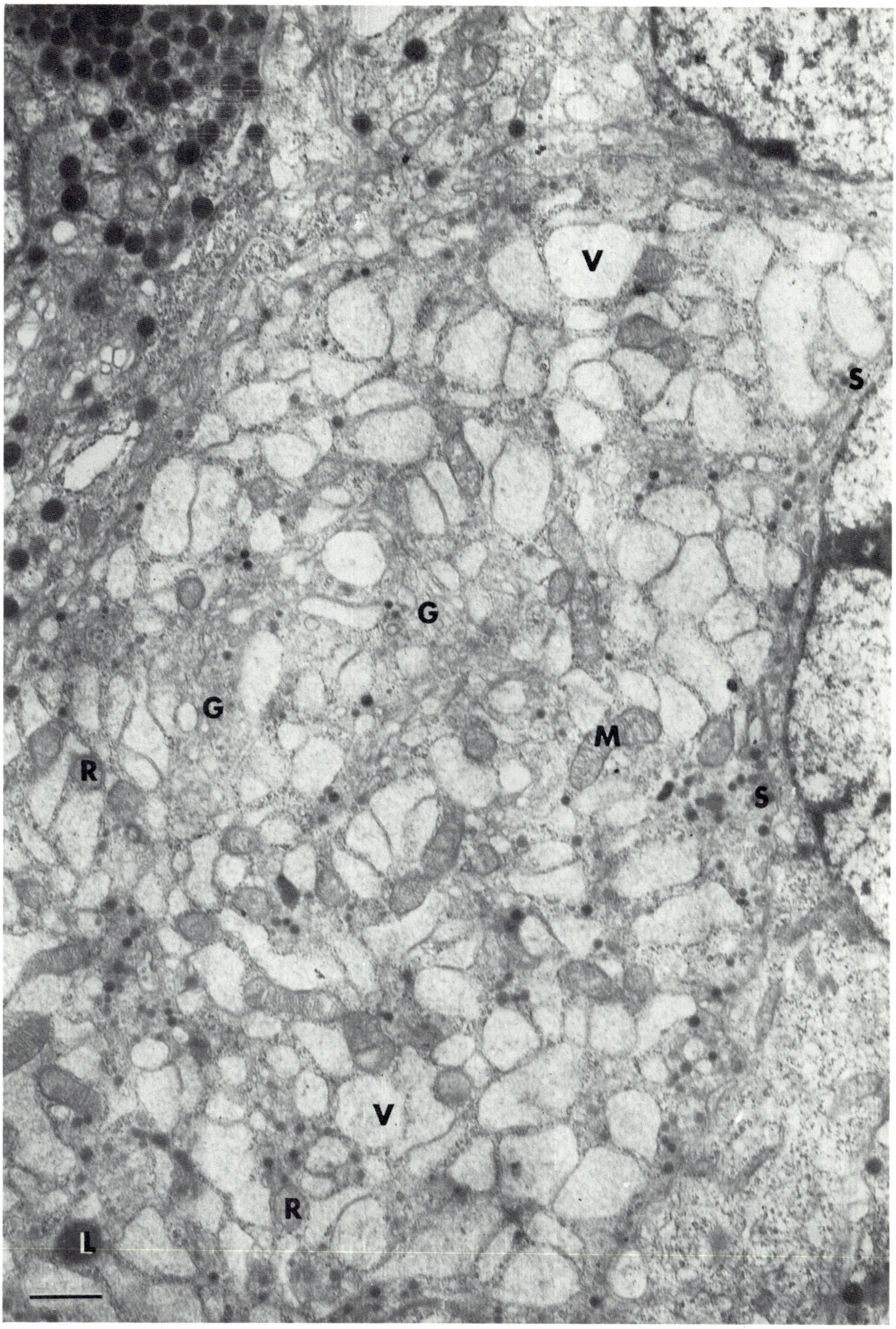

Fig. 3.3. A thyroidectomy cell after PTU treatment. The thyroidectomy cell contains a vacuolar endoplasmic reticulum (V), lysosomes (L), and some granules (S) along the periphery and within the cell. Round mitochondria (R) appear to predominate over the rod shaped (M). Golgi complexes (G) are shown in the center of the cell. × 12,000. (Veronal acetate buffered osmic acid)

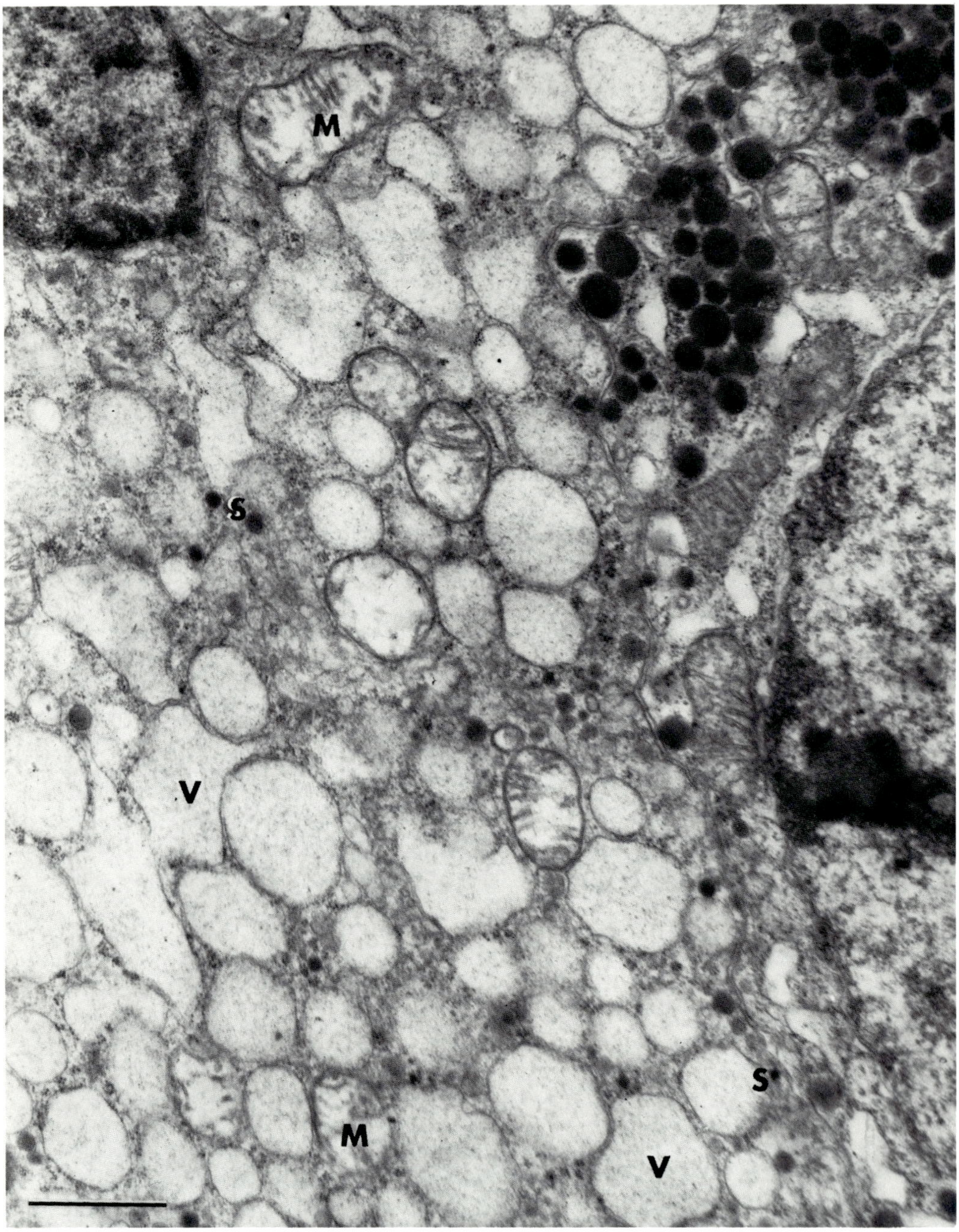

Fig. 3.4. A section of a greatly enlarged thyrotrope after PTU treatment. A part of a somatotrope is shown in the upper right hand corner of the micrograph. Few granules (S) are shown. Some mitochondria (M) are swollen with obliterated cristae. The endoplasmic reticulum is greatly vacuolated (V). × 20,000. (Veronal acetate buffered osmic acid)

lesser number of thyroidectomy cells. On the basis of this evidence it appears that estrogen prevents the dramatic change in TSH cells after PTU treatment.

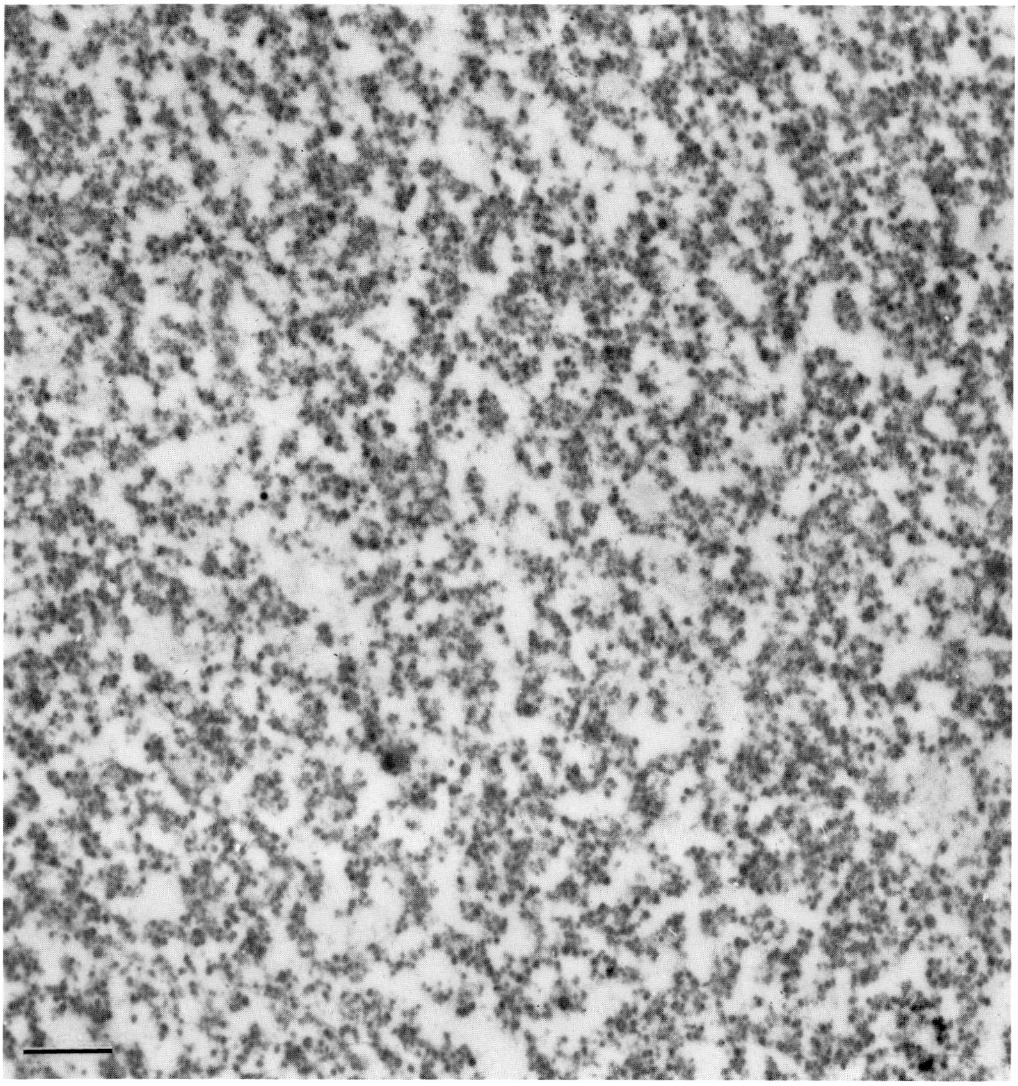

Fig. 35. A section of the D1 HSP pellet. This granule fraction is essentially free of microsomal contamination and consists of secretory granules 150 mμ in maximum diameter. $\times$ 12,000. (Veronal acetate buffered osmic acid)

After thyroidectomy or PTU treatment there is an increase in thyroidectomy cells over TSH cells in the intact group of animals (Vanha-Perttula, 1966). Since mitosis is rarely found in pituitary cells, what accounts for the increase in TSH cells? It is suggested that chromophobe-stem cells are transformed into thyroidectomy cells. This follows the postulations of Severinghaus (1937). Dingemans (1969) conducted thyroidectomy studies using mice, which included pituitary cell counts before and after extirpation. He found that the number of LTH, STH, and LH cells was reduced after thyroidectomy and suggests that these cells may give rise to thyroidectomy cells.

Thyrotropes did not change after adrenalectomy or castration, but TSH granules increased in number after α-ethyltryptamine administration. Administration of adrenal cortical blocking agents or corticoids caused a slight reduction in number of thyrotropes. Since after thyroidectomy or propylthiouracil treatment only TSH cells are greatly enlarged, it is evident that the cell type which produces TSH has been identified.

TSH CONTENT OF PITUITARY GRANULE FRACTIONS

Eighty-eight percent of the original TSH activity of the FS1 fraction layered on the continuous gradient was recovered in the gradient zones. It was found that 83% of the recovered TSH activity was associated with the D1HSP granule pellet (Fig. 3.5). Four percent of the remaining activity recovered from the gradient was found in solution in the SA zone and 1.5% was associated with the FD2HSP pellet. The B and C pellets, containing microsomal and ribosomal material, had trace amounts of TSH activity. This may mean that in removing zone C a small amount of the D1 zone was included.

An electron micrograph (Fig. 3.5) of the D1HSP pellet is shown. It is free of contamination of other cell organelles and had granules with a mean diameter of 89 mμ. This compares favorably with the granule measurements from sections of TSH cells from whole tissue in which a mean of 85 mμ was observed. These granule measurements and bioassays provide additional support that these cells are thyrotropes.

In summary, the identification of TSH cells is similar to those given by Farquhar and Rinehart (1954), Lever and Peterson (1960), Kurosumi and Oota (1966), and the recent findings of Nakane (1968).

Nakane (1970) and Baker and Yu (1971), using the peroxidase-labeled antibody method, found the distribution and characteristics of the TSH cell type to be similar to those that have been described.

REFERENCES

Baker, B. L., and Yu, Y.-Y. (1971). The thyrotropic cell of the rat hypophysis as studied with peroxidase-labeled antibody. *Amer. J. Anat.* **131**, 55.

Barnes, B. G. (1963). *In* "Cytologie de l'Adenohypophyse" (J. Benoit and C. Da Lage, eds.) p. 91. Editions du Centre National de La Recherche Scientifique, Paris.

Bogdanove, E. M., and Halmi, N. S. (1953). Effects of hypothalamic lesions and subsequent PTU treatment on structure of anterior pituitary gland. *Endocrinology* **53**, 274.

Catchpole, H. R. (1949). Distribution of glycoprotein hormones in the anterior pituitary gland of the rat. *J. Endocrinol.* **6**, 218.

Contopoulos, A. N., Simpson, M. E., and Koneff, A. A. (1958). Pituitary function in the thyroidectomized rat. *Endocrinology* **63**, 642.

D'Angelo, S. A. (1961). TSH rebound phenomenon in the rat adenohypophysis. *Endocrinology* **69**, 834.

Deminatti, M. (1962). Comparative histoautoradiographical study of the anterior pituitary gland of *Carassium auratus* after administration of S35 labeled Na_2SO_4 and S35 labeled methionine. *C. R. Acad. Sci.* **254**, 1510.

Dhom, G., and Tietze, H. V. (1962). Zur experimentellen Histophysiologie der Beta-zellen der Rattenhypophyse nach Untersuchungen mit der Perameisenaure-Alcian Blau-Reaktion. *Endokrinologie* **42**, 284.

Dingemans, K. P. (1969). On the origin of thyroidectomy cells. *J. Ultrastruct. Res.* **26**, 480.

Ezren, C., and Murray, S. (1963). *In* "Cytologie de l'Adenohypophyse" (J. Benoit and C. Da Lage, eds.), p. 183. Editions du Centre National de La Recherche Scientifique, Paris.

Fand, S. B., and Thorell, B. (1962). Spectrophotometry of the periodic acid-Schiff reaction with pituitary hormones *in vitro* and in histological sections. *J. Cell Biol.* **13**, 239.

Farquhar, M. G., and Furth, J. (1959). Electron microscopy of experimental pituitary tumors. *Amer. J. Pathol.* **35**, 698.

Farquhar, M. G., and Rinehart, J. F. (1954). Cytologic alterations in the anterior pituitary gland following thyroidectomy: An electron microscopic study. *Endocrinology* **55**, 857.

Feltkamp, C. A., and Kwa, H. G. (1965). Transformation of pituitary tumours from prolactin into TSH-cell types. 1. Electron microscopic observations. *Acta Endocrinol. (Copenhagen), Suppl.* **100**, 161. (abstr.) (1950).

Gomori, G. (1950). Aldehyde fuchsin: A new stain for elastic tissue. *Amer. J. Clin. Pathol.* **20**, 665.

Greenspan, F. S., and Haragadine, J. R. (1965). The cellular localization of pituitary thyrotropic hormone. *J. Cell Biol.* **26**, 177.

Griesbach, W. E. (1941). Studies on experimental goitre. II. Changes in the anterior pituitary of the rat produced by Brassica seed diet. *Brit. J. Exp. Pathol.* **22**, 245.

Griesbach, W. E., and Purves, H. D. (1945). The significance of the basophil changes in the pituitary accompanying various forms of thyroxine deficiency. *Brit. J. Exp. Pathol.* **26**, 13.

Guyer, M. F., and Claus, E. G. (1937). Vacuolation of anterior pituitary gland following castration, implantation of cancer tissue and thyroidectomy. *Anat. Rec.* **67**, 145.

Halmi, N. S. (1950). Two types of basophils in the rat anterior pituitary gland and their respective cytophysiological significance. *Endocrinology* **47**, 289.

Halmi, N. S. (1952). Two types of basophils in the rat pituitary: "Thyrotrophs" and "gonadotrophs" vs. beta and delta cells. *Endocrinology* **50**, 140.

Herlant, M. (1943). Recherches sur la localisation histologique des hormones gonadotropes femelles au niveau de l'hypophyse antérieure. *Arch. Biol.* **54**, 225.

Herlant, M. (1960). Etudé critique de deux technique nouvelles destinées a mettre en evidence les differeétes catégories cellulaires présentes dans la glande pituitaire. *Bull. Microsc. Appl.* **10**, 37.

Herlant, M. (1964). The cells of the adenohypophysis and their functional significance. *Int. Rev. Cytol.* **17**, 299.

Hohlweg, W., and Junkman, K. (1933). Über die Beziehungen Zwischen Hypophysenvorderlappen und Schild druse. *Pfluegers Arch. Gesamte Physiol. Menschen Tiere* **232**, 148.

Hotchkiss, T. D. (1948). A microchemical reaction resulting in the staining of polysaccharide structures in fixed tissue preparations. *Arch. Biochem.* **16**, 131.

Iino, A., Yamada, T., and Greer, M. (1961). Effect of graded doses of propylthiouracil on biosynthesis of thyroid hormones. *Endocrinology* **68**, 582.

Kurosumi, K., and Oota, Y. (1966). Corticotrophs in the anterior pituitary glands of gonadectomized and thyroidectomized rats as revealed by electron microscopy. *Endocrinology* **79**, 808.

Lever, J. D., and Peterson, R. R. (1960). Cellular identities in the rat pituitary. *Trans. N.Y. Acad. Sci.* [2] **22**, 504.

McManus, J. F. A. (1946). Histological demonstration of mucin after periodic acid. *Nature (London)* **158**, 202.

Messier, B. (1965). Demonstration of TSH in thyroidectomized mice. *Union Med. Can.* **94**, 88.

Nakane, P. K. (1968). Simultaneous localization of multiple tissue antigens using the peroxidase-labeled antibody method: A study on pituitary glands of the rat. *J. Histochem. Cytochem.* **16**, 557.

Nakane, P. K. (1970). Classifications of anterior pituitary cell types with immunoenzyme histochemistry. *J. Histochem. Cytochem.* **18**, 9.

Paget, G. E., and Eccleston, E. (1960). Simultaneous specific demonstration of thyrotroph, gonadotroph and acidophil cells in the anterior hypophysis. *Stain Technol.* **35**, 119.

Pasteels, J. L., and Herlant, M. (1963). Notions nouvelles sur la cytologie de l'antehypophyse chèz le rat. *Z. Zellforsch. Mikrosk. Anat.* **56**, 20.

Pearse, A. G. E. (1948). Cytochemistry of gonadotropic hormones. *Nature (London)* **162**, 651.

Pearse, A. G. E. (1949). The cytochemical demonstration of gonadotropic hormone in the human hypophysis. *J. Pathol. Bacteriol.* **61**, 195.

Pearse, A. G. E. (1950). Differential stain for the human and animal anterior hypothysis. *Stain Technol.* **25**, 95.

Potvliege, P. R. (1970). Effects of estrogen on pituitary morphology in goitrogen treated rats: An electron microscopic study. *Anat. Rec.* **160**, 595.

Purves, H. D. (1956). Changes in basophil cells in the rat anterior pituitary after thyroidectomy. *J. Endocrinol.* **13**, 365.

Purves, H. D., and Griesbach, W. E. (1951a). The site of thyrotrophin and gonadotrophin production in the rat pituitary studied by McManus-Hotchkiss staining for glycoprotein. *Endocrinology* **49**, 244.

Purves, H. D., and Griesbach, W. E. (1951b). Specific staining of the thyrotropic cells of the rat pituitary by the Gomori stain. *Endocrinology* **49,** 427.

Richards, J. B., and Ingbar, S. H. (1959). The effects of prophylthiouracil and perchlorate on the biogenesis of thyroid hormone. *Endocrinology* **65,** 198.

Severinghaus, A. E. (1935). A suggestive correlation of cytological changes with secretory activity in the cells of the normal anterior pituitary gland. *Anat. Rec.* **61,** Suppl. 61.

Severinghaus, A. E. (1937). Cellular changes in the anterior hypophysis with special reference to its secretory activities. *Physiol. Rev.* **17,** 556.

Severinghaus, A. E., Smelser, G. K., and Clark, H. M. (1934). Anterior pituitary changes in the adult male rat following thyroidectomy. *Proc. Soc. Exp. Biol. Med.* **31,** 1127.

Slingeiland, D. W., Graham, D. E., Josephs, R. K., Mulvey, P. F., Trakas, A. P., and Yamazaki, E. (1959). The effect of propylthiouracil on the conversion of monoiodotyrosine to diiodotyrosine. *Endocrinology* **65,** 178.

Theret, C., and Renault, H. (1964). Morphocineses différentes de tumeurs thyréotropes aprés radiothyroidectomie et addition de cancerigens chimiques. *Bull. Cancer* 51, 505.

Vanha-Perttula, T. (1966). Esterases of the rat adenohypophysis. *Acta Physiol. Scand.* **69,** Supp. 283. 104 pp.

Zechwer. I. T., Davison, L. W., Keller, T. B., and Livingood, C. S. (1935). Structural changes in the pituitaries of thyroidectomized rats. *Amer. J. Med. Sci.* **190,** 145.

Chapter 4

Gonadotropes

The two gonadotropic hormones are grouped together for the study of their sites of production because they are closely associated physiologically and it has been difficult to identify which cell produces each hormone.

Cytological studies of pituitary glands were made during castration and pregnancy long before gonadotropins were known. Comte (1898) reported that during pregnancy the proportions of acidophils, basophils, and chromophobes change. Fichera (1905) described, for the first time, that changes occur in the pituitary after castration, and that the changes were reversed by injections of extracts of the gonads. Schleidt (1914) observed many vacuolated basophils following castration in the human and designated these as "signet ring" cells. Addison (1917) also described castration cells in the rat.

Evans and Simpson (1929) and Engle (1929) reported that gonadotropic activity was elevated after castration. Engle (1929) linked this increased gonadotropic activity with basophils. The importance of the hypophysis in maintaining gonadal function was convincingly demonstrated by the data obtained by Smith (1930) using hypophysectomized rats. After hypophysectomy the gonads atrophied but were restored to normal when pituitary extracts were injected or implants of the gland were made.

After Smith's work became known, attempts were made by other workers to separate the gonadotropins from other hormones of the anterior pituitary gland. Fevold *et al.* (1931) were the first to accomplish partial separation of FSH and LH. Fevold *et al.* (1931, 1933) studied the biological effects of these gonadotropic preparations in rats and rabbits. Additional descriptions of the effects produced by follicle stimulating hormone fractions were reported by Chow (1943) and Fraenkel-Conrat *et al.* (1943). The action of these two hormones was to promote the growth of ovarian follicles and affect the germinal epithelium of the testes. Greep *et al.* (1941) and Fevold (1943) reported that luteinizing hormone stimulated luteinization of the ovaries and maintained the interstitial cells of the testes.

LIGHT MICROSCOPY OF GONADOTROPES

Guyer and Claus (1937) described the vacuolation that occurred in certain basophils after castration. Severinghaus (1937) postulated that basophils secrete FSH and acidophils LH. Smelser (1944) and Giroud and Martinet (1948) corrected the erroneous hypothesis of Severinghaus by administering extracts of acidophilic and basophilic areas of ox and pig pituitaries to immature rats and mice and observed that both FSH and LH effects were elicited by extracts of basophilic parts of the pituitary glands. Wolfe and Brown (1942) found that certain basophils increased in size and number following castration.

Romeis (1940) claims to have distinguished FSH and LH cells using Kresazan staining. Employing the metachromatic technique, Herlant (1943) concluded that basophils are the source of gonadotropic hormones. Basophils have also been implicated as the source of FSH, LH, and TSH by Pearse (1948, 1949, 1950, 1953), Catchpole (1949), and Hotchkiss (1948), using the periodic acid Schiff (PAS) staining technique. Two types of gonadotropes were distinguished by Halmi (1950, 1952), and Purves and Griesbach (1951) clearly distinguished between two types of basophils in rat glands using PAS stain.

FSH and LH Gonadotropes Distinguished by Staining Methods

Purves and Griesbach (1951) described gonadotropes as oval cells which stain with PAS and are located on the dorsal and ventral parts of the anterior pituitary gland. The PAS reaction was correlated with the gonadotropic hormone content of the gland. These cells were involuted after estrogen treatment and became hypertrophied as a result of castration.

Using PAS stain Purves and Griesbach (1952) described two gonadotropes. The central gonadotrope which stains red is prominent just before the onset of sexual maturity in the female rat and degranulates just before ovulation. During pregnancy and estrogen injection these central gonadotropes degranulate, hypertrophy, and show morphological evidence of increased activity. After castration these cells showed a marked accumulation of glycoprotein, and assays of the glands made at this time showed them to have a high LH content. On the basis of this evidence, it was concluded that the central gonadotropes secrete LH.

These workers also defined a population of purple staining basophils as peripheral gonadotropes. Estrogen injection caused an increase of hormone in the gland and increased granulation of these cells. During pregnancy they also exhibited increased granulation. Peripheral basophils were well granulated during early castration and assays conducted at this time showed high FSH activity, indicating that these peripheral gonadotropes secrete FSH.

Purves and Griesbach (1954) presented more data to support their views on gonadotropic cells. Administration of testosterone propionate caused an increase in FSH and a decrease in LH content of rat pituitaries. They correlated this elevated pituitary FSH with an increase in granulation in peripheral gonadotropes, and the decreased LH with a reduced number of granules in the central gonadotropes. This work agreed with the earlier findings of Greep and Jones (1950).

Barnett *et al.* (1956) reported that a 2.5% solution of tricholoracetic acid extracted PAS positive FSH and TSH but not LH as shown by assays. The authors believed this to be a useful histochemical method to demonstrate LH secreting cells. They reported that LH activity was distributed throughout the gland in contrast to the earlier report of Purves and Griesbach (1954) who stated that LH cells were in the central part of the pituitary. Rennels (1957, 1963), Hildebrand *et al.* (1957), Hellbaum *et al.* (1961), and Chowdhury *et al.* (1971) also questioned the findings of Purves and Griesbach. Their evidence indicated that the central PAS purple cells produce FSH, and the peripheral PAS red gonadotropes produce LH. This controversy about the staining affinities of LH and FSH gonadotropes has not been completely resolved. Herlant (1964), Hartley (1959), Kracht (1957), and Vanha-Perttula (1966) are among the investigators who agree with the postulations of Purves and Griesbach (1954). Physiological and electron microscopic data seem to support the Purves and Griesbach postulation.

A study of the rat hypophysis by Herlant (1960) utilized an alcian blue–PAS stain. He found that a pH 0.2 LH cells stain red and FSH cells violet. Vanha-Perttula (1966) followed the same procedure but used a pH of 3.0. Under these conditions LH cells stain light green and FSH cells violet. He also used an aldehyde–thionin staining technique which indicated that LH cells stain red and FSH cells light blue.

Distinguishing FSH and LH Cells by Immunofluorescence Method and Autoradiography

Immunofluorescence techniques have been employed to demonstrate FSH and LH cells. Midgley (1963), Robyn *et al.* (1964), and Bain and Ezrin (1970) used this histochemical means to determine LH cells in man. Della Corte and Biondi (1964) used immunofluorescence methods to label FSH cells in pig pituitary glands. Koffler and Fogel (1964) and Fogel and Koffler (1964) used purified antisera to human chorionic gonadotropin and human menopausal gonadotropin specific for FSH and LH, respectively, to characterize LH and FSH cells in the human pituitary. Recently, luteinizing hormone producing cells have been identified by Pomerantz and Simmons (1968) and Herlant and Ectors (1970) in cow and pig pituitaries using immunofluorescence methods.

Autoradiographic techniques have been used to distinguish gonadotropes from other cells. Ducommun (1965) found that tritiated proline was incorporated into gonadotropes of castrate animals; similarly, more tritiated proline was incorporated into TSH cells after thyroidectomy. These data indicate that the gonadotropes are distinct from thyrotropes. Kobayashi *et al.* (1967) demonstrated that ^{3}H-leucine is incorporated into the gonadotropes of castrate and estrogen-treated rats.

Changes in FSH and LH Cells During the Estrous Cycle

Barnes (1962a) traced the ultrastructure of the mouse pituitary gland through the estrous cycle. She designated the cells that were degranulated early in proestrus as FSH and those which degranulated later as LH cells. Herlant (1964) traced gonadotropes through the estrous cycle in the mole. Roos (1968) traced gonadotropic cells through the estrous cycle in the rat. FSH cells were degranulated during the height of diestrus, which is related to the wave of follicular growth that takes place at this time of the cycle. The FSH cells were also degranulated during the afternoon of proestrus. Degranulation of LH cells occurred during the morning of proestrus and reached a peak during the critical period of proestrus at 4:00 PM. The second surge of FSH occurs during the critical period at the same time as LH. Roos also found LH cells to be centrally located, thereby agreeing with the work of Purves and Griesbach (1954). Monroe *et al.* (1969) used radioimmunoassays to measure LH levels in the rat during the estrous cycle. They found serum LH levels to be 25 times greater during the afternoon of early proestrus. These data seem to support the findings of Roos (1968).

ULTRASTRUCTURE OF GONADOTROPES

Electron microscopic studies of gonadotropes are numerous as indicated by the studies of Farquhar and Rinehart (1954) and Yoshimura and Harumiya (1965) in the rat; Yamada and Sano (1960) and Barnes (1963) in the mouse; and Girod and Dubois (1965) and Dekker (1967) in the hamster. Only the study by Barnes (1963) has distinguished between FSH and LH cells.

The earliest electron microscopic study of gonadotropes and the changes they undergo after castration was made by Farquhar and Rinehart (1954) and Farquhar (1955). They described FSH gonadotropes as large, round cells located on capillaries. They noted that these cells undergo vacuolation and that the Golgi apparatus enlarges 6 days after castration. The smaller LH cells are found on capillaries and appear filagreed. These cells undergo castration changes 60 days after gonadectomy. Their report did not provide enough definitive data to distinguish the two cell types and it has been suggested by Kurosumi and Oota (1966) that the cells described by Farquhar and Rinehart are two different forms of the same cell.

Farquhar and Rinehart (1954) and Yoshimura and Harumiya (1965) in their long-term castration studies estimated granules in the gonadotropes to be about 200 mμ in diameter and less electron dense than granules in acidophils. The latter workers have shown FSH granules to be smaller, 150–200 mμ, and LH granules larger, 200–250 mμ in diameter. Fractionation of rat pituitary granules by Hartley *et al.* (1960), Hymer and McShan (1963), and Perdue and McShan (1962) have shown that gonadotropic activity was associated with granules 150–200 mμ in diameter.

In recent electron microscopic study by Kurosumi and Oota (1968) it was found in persistent estrus rats that LH cells were well granulated, indicating a storage of LH, but FSH cells were sparcely granulated. LH cells in persistent diestrus rats were slightly activated and FSH cells were atrophic. Prior to 1968 the identification of these two gonadotropes was inconclusive, but now by use of newer research techniques they have been more positively identified.

Follicle-Stimulating Gonadotropes

In our studies we found that the follicle-stimulating gonadotrope is the largest cell found in the male rat pituitary gland (Fig. 4.1). In the female it is smaller (Fig. 4.2) and not as abundant. This cell type was abundant

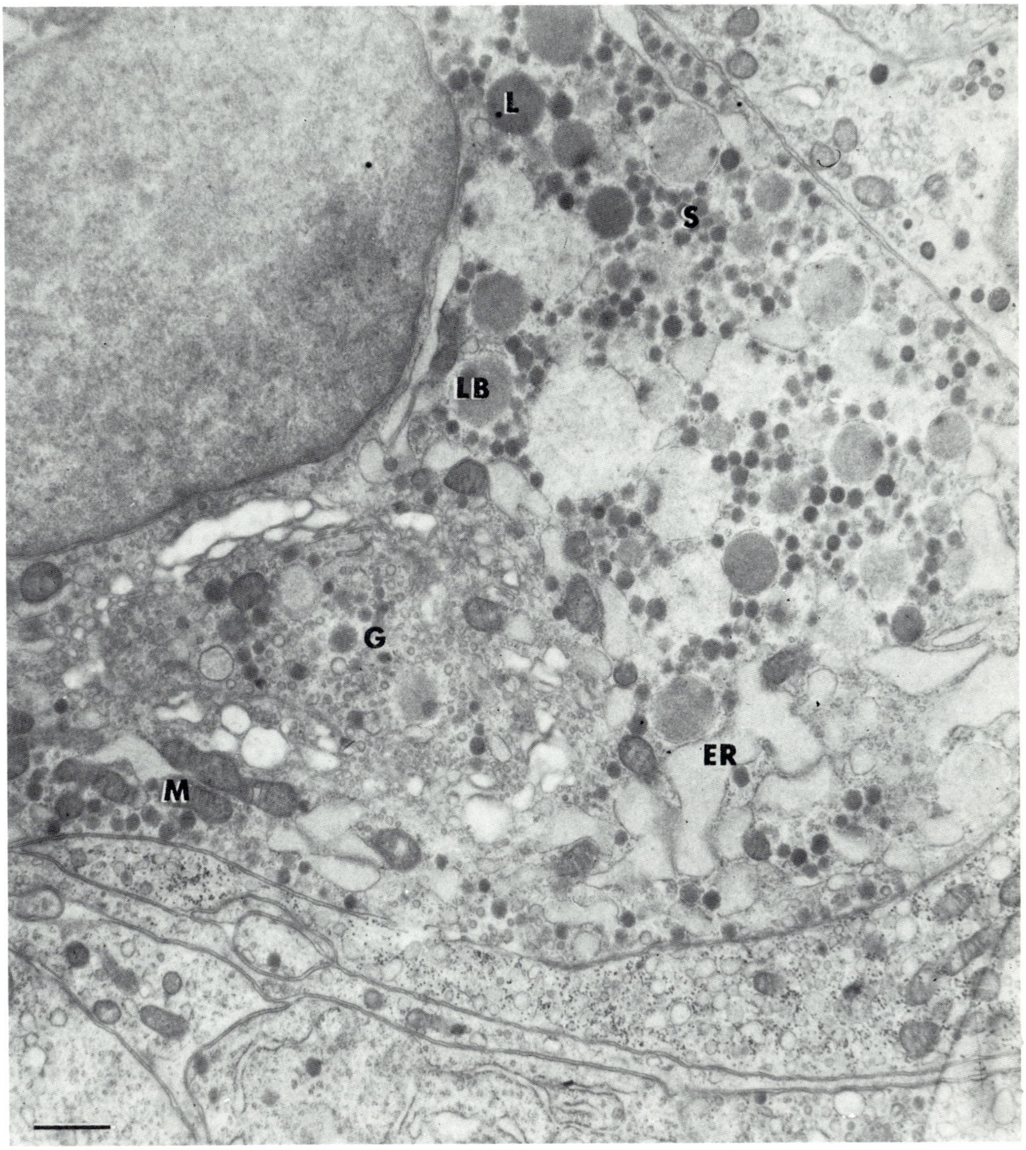

Fig. 4.1. A FSH gonadotrope from a male rat pituitary. This very large cell contains secretory granules (S) of about 200 mμ in diameter and large light staining bodies (LB) up to 1.0 μ in diameter. These organelles can be seen in various stages of formation in the Golgi complex (G). Mitochondria (M) with dense cristae, lysosomes (L), and several isolated areas of endoplasmic reticulum (ER) are shown. $\times$ 10,000. (Veronal acetate buffered osmic acid)

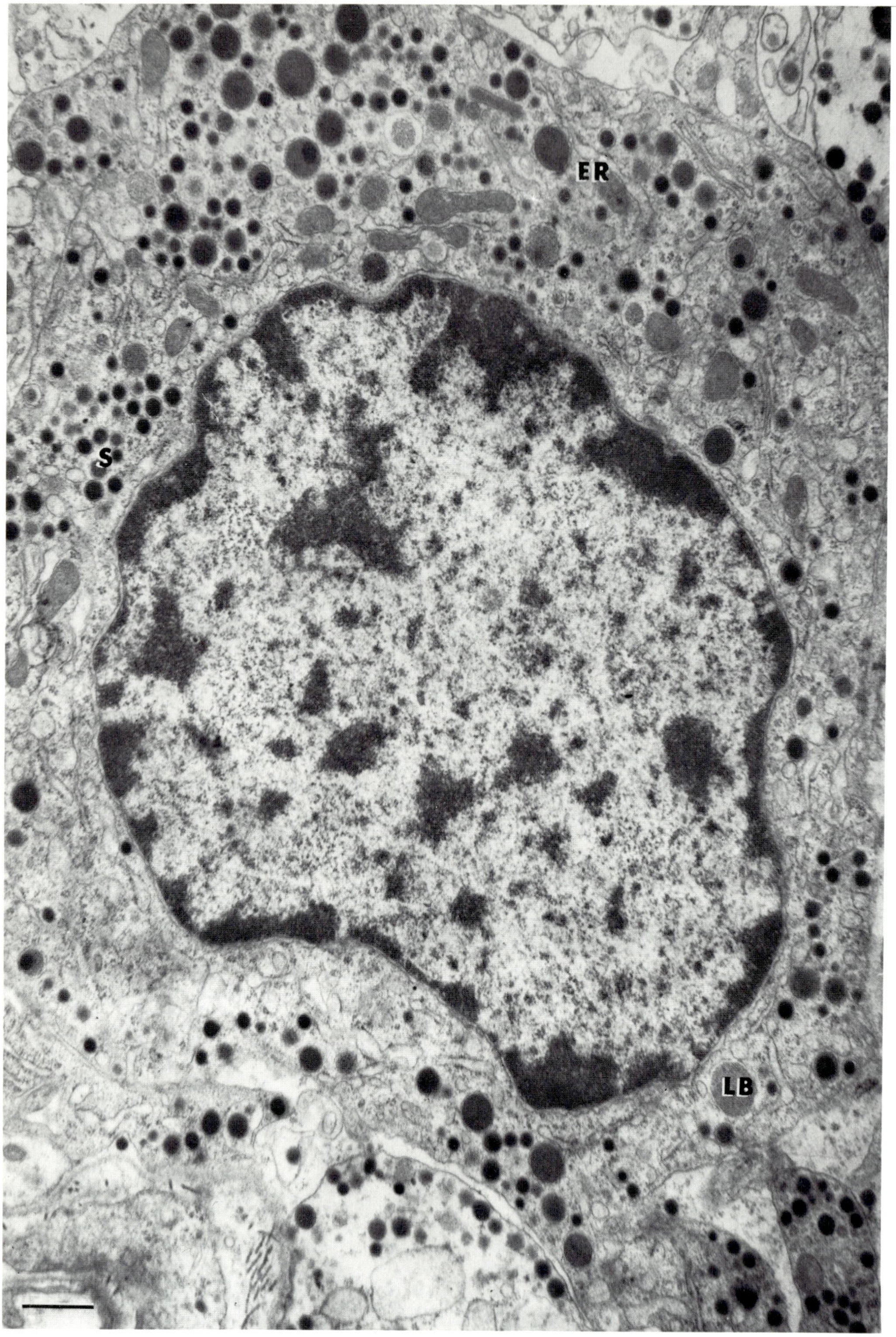

Fig. 4.2. A FSH gonadotrope in a female rat pituitary. This cell is smaller in the female than in the male but it still contains small secretory granules (S) and large bodies (LB) of varying densities. The endoplasmic reticulum (ER) is scattered and the mitochondria are short rods or circular in shape. × 11,200. (Glutaraldehyde–paraformaldehyde)

in most sections of male rat pituitary glands. The FSH cell is the second most frequently observed type found in the male rat pituitary, with perhaps 30% of all the cells secreting FSH; in the female it may be about 10%.

These cells are round and usually found on a capillary. The nucleus is round or indented on one side. The granules are small and vary in density and are distributed throughout the cell. The granule diameter ranges from 75–200 mμ with a mean of 126 mμ. There are also large amorphous bodies throughout the cells which are unique to the FSH cell (Figure 4.1). These opaque bodies are not always seen in FSH cells (Fig. 4.3). They have been observed by others (Cardell, 1961; Farquhar and Rinehart, 1954; Kurosumi and Oota, 1968). These light-staining bodies range from 0.7–1.2 μ in diameter and can be seen being formed along with the granules in the Golgi complex (Fig. 4.4). Depending upon the fixation these large bodies may be as electron dense as the secretory granules.

There are several Golgi areas encircling the nucleus in FSH cells, and they usually appear hypertrophied and possess a well-developed network of dilated sacs and small vesicles. The mitochondria of FSH cells are filamentous rods of various shapes usually exhibiting a dense matrix. Cristae are more continuous and parallel to one another as compared to those observed in other pituitary cell types.

The endoplasmic reticulum is not as well developed in the FSH gonadotrope as in the acidophils but is more extensive than in TSH cells. In hypertrophied FSH cells the endoplasmic reticulum is more irregular and has dilated sacs. In an inactive cell the endoplasmic reticulum is more vesicular with few lamellae, and ribosomes are not always attached to the membranes. Lysosomes, multivesicular bodies, centrioles, and occasionally a cilium are also observed.

Luteinizing Hormone Gonadotropes

Luteinizing hormone cells are more common in the female rat, constituting as much as 20% of the total cells, than in the male, where there are about 5–10% of these cells. They are more usually found anteromedially in the pituitary. The LH cells are larger than thyrotropes (Fig. 4.5) and are often located on a capillary. They are usually polygonal in shape with eccentrically positioned nuclei. The secretory granules of these gonadotropes are more electron dense than those of FSH cells but less dense than those of acidophils. Sections of luteinizing hormone granules average 145 mμ in diameter with a range of 75–235 mμ. This gona-

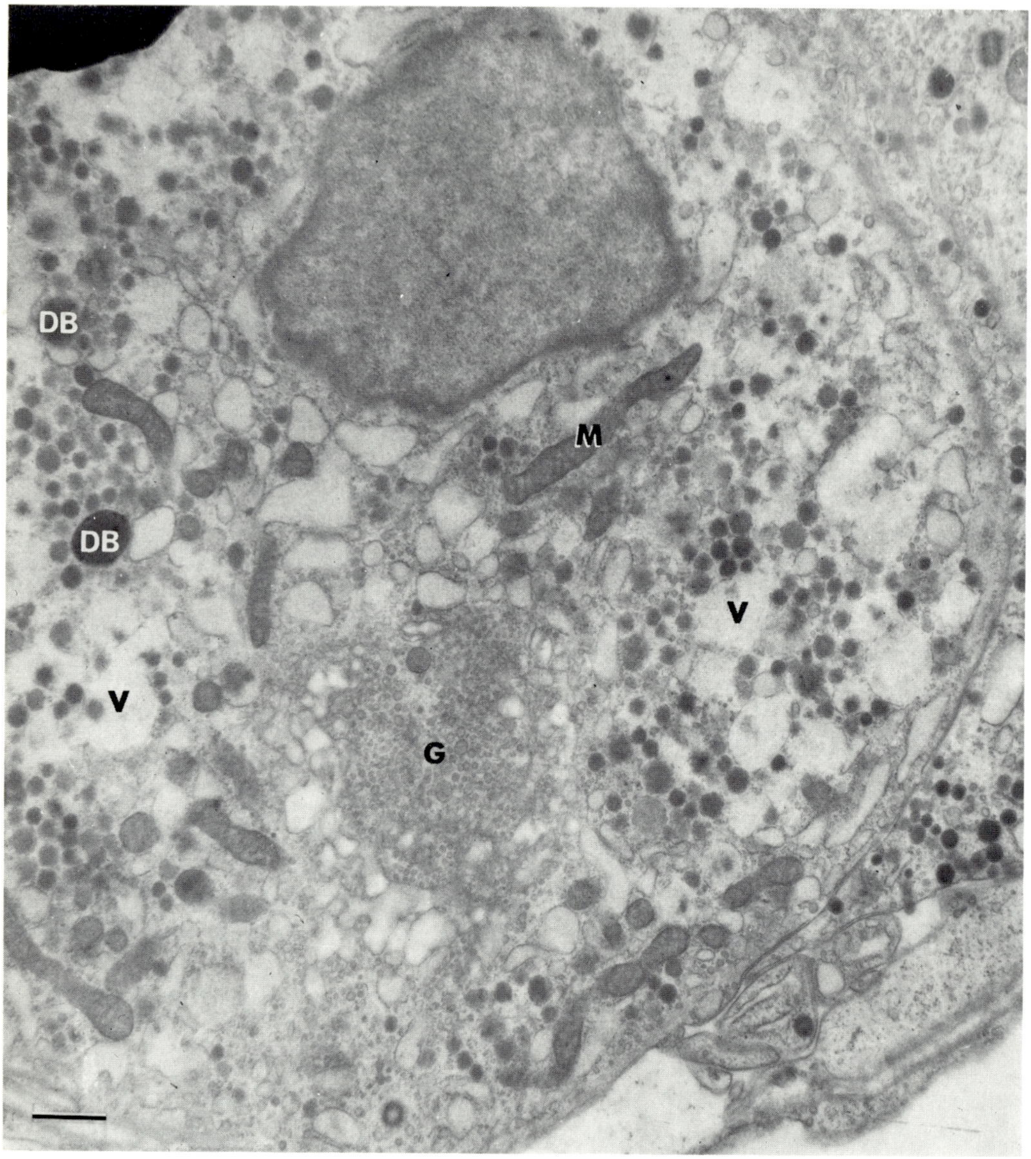

Fig. 43. A FSH cell from a male rat pituitary. Two large dense bodies (DB), probably lysosomes, are shown. But the large opaque bodies are not present and the cell, although well granulated, appears vacuolated (V). An enlarged Golgi complex (G) and long mitochondria (M) are present. × 10,000. (Veronal acetate buffered osmic acid)

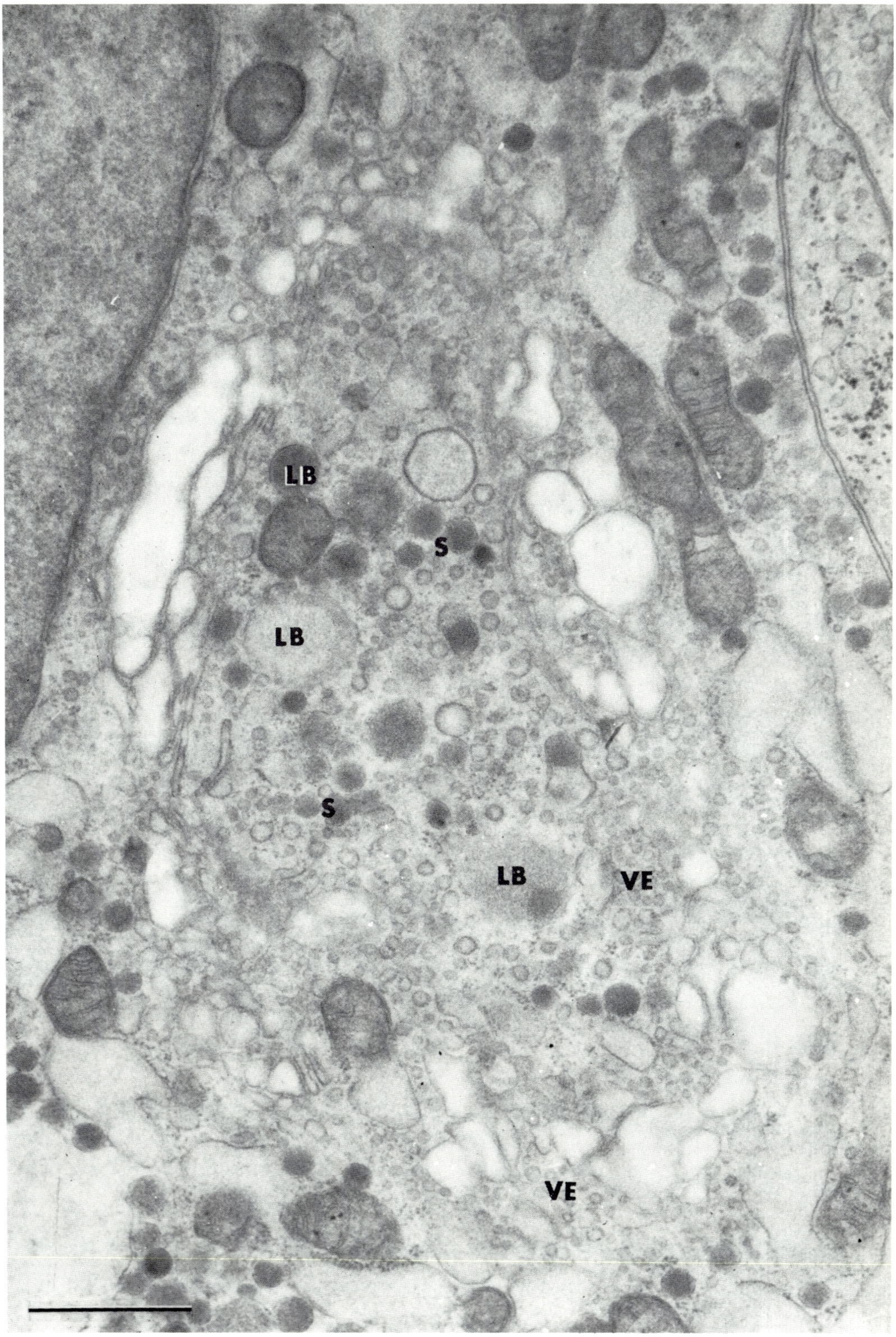

Fig 4.4. In this micrograph a Golgi complex is shown in a FSH cell. Vesicles (VE), secretory granules (S) in various stages of formation, and opaque light bodies (LB) are found in the Golgi complex. × 25,000. (Veronal acetate buffered osmic acid)

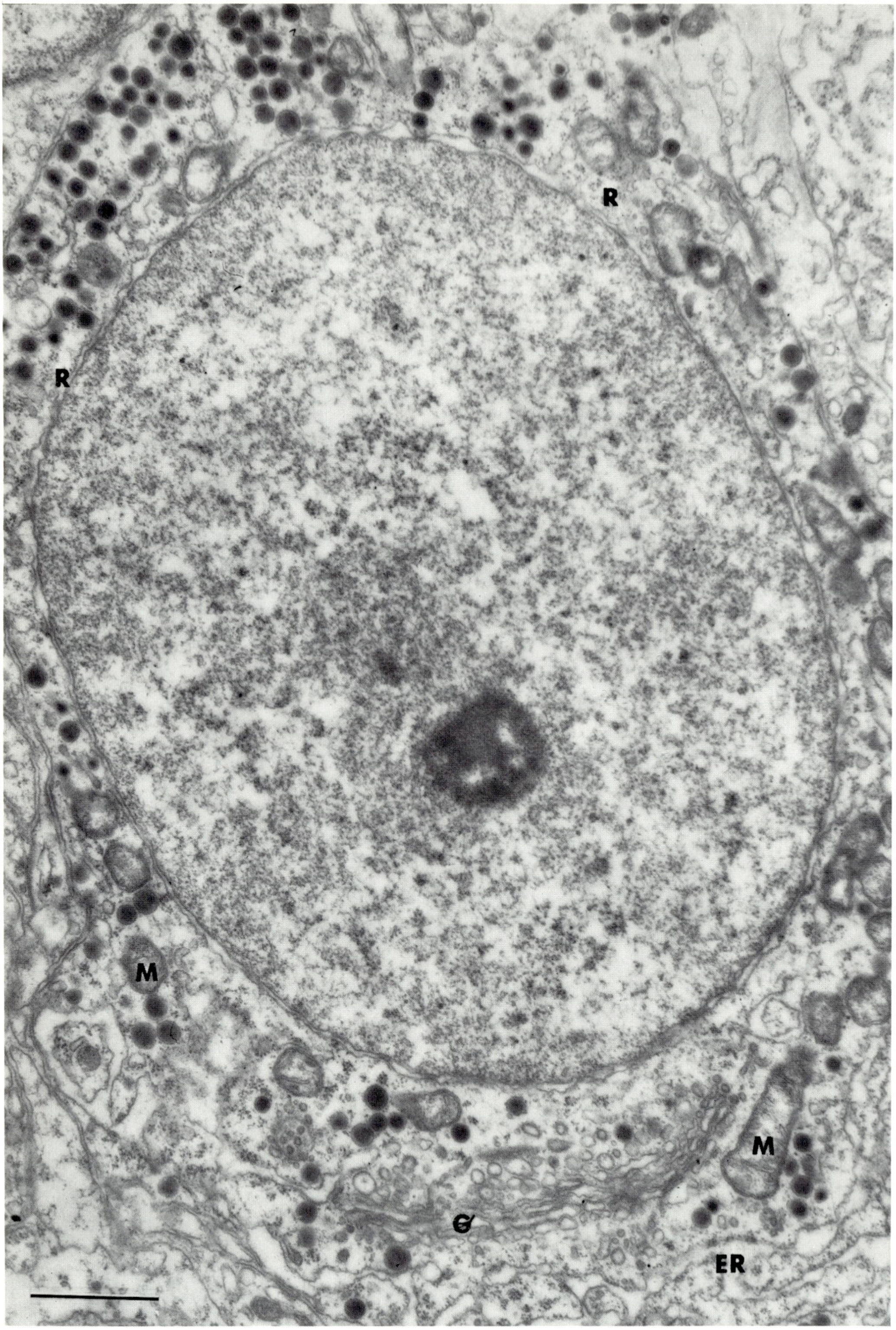

Fig. 4.5. A luteinizing hormone (LH), gonadotrope from a male rat anterior pituitary gland. This medium-sized cell contains an extensive Golgi complex (G). Mitochondria (M) are round or rodlike with sparse interrupted cristae. Several free ribosomes (R) and flattened saccules of endoplasmic reticulum (ER) with and without attached ribosomes are also shown. $\times$ 20,000. (Veronal acetate buffered osmic acid)

dotrope does not contain the large amorphous bodies that are present in the FSH cells.

When the LH cells are in an inactive state, the Golgi apparatus and endoplasmic reticulum are rather inconspicuous and poorly developed. In actively stimulated cells there are extensive Golgi areas (Fig. 4.6) with granules in different stages of formation. However, even in the stimulated LH cell the Golgi complex is less extensive than in the FSH cell. The endoplasmic reticulum consists of scattered vesicular areas. These membranes are intermittently dotted with ribosomes but some are free in the cytoplasm. Both short, rodlike and round mitochondria are found in these cells. Few lysosomes, cilia, or centrioles are present.

Effect of Castration on the Fine Structure of the Gonadotropes

Rats were killed 7, 14, 30, and 60 days following castration, and the cytological changes in the pituitary were studied. Castration changes appear to be more dramatic in the male rat. Cytological changes occurring in the male rat pituitary are described as follows.

Seven days after gonadectomy the most noticeable changes that occurred were in the gonadotropes. The FSH cells showed an increase in granulation but were devoid of the large amorphous bodies which were usually found in intact male FSH cells (Fig. 4.7). There was an increase, however, in lysosomes. Whether the large amorphous bodies are antecedents of lysosomes or are engulfed or absorbed by the latter is not clear. The nucleus of the FSH cells was irregular and the cytoplasm was more vacuolated but the changes were not as great as those found in LH cells.

Luteinizing hormone cells, rather inconspicuous in the intact male rat, were enlarged and apparently increased in number by 7 days after castration. The hypertrophied cells had a well-developed endoplasmic reticulum, an enlarged Golgi apparatus and many granules (Fig. 4.8). In some cells the vesicular endoplasmic reticulum was dilated and enlarged and there was colloidal material in some of the vacuoles of the endoplasmic reticulum.

Between 14 (Fig. 4.9) and 60 days following castration the FSH cells became more vacuolated and the number of lysosomes increased. The endoplasmic reticulum became smooth, and the Golgi apparatus became more distinct; the number of granules decreased (Fig. 4.10). The mitochondria changed from the filamentous to the short rodlike type.

The LH cell hypertrophied greatly after long-term castration. There was an increase in granulation up to 14 days after castration (Fig. 4.11), and thereafter granulation tended to decrease. The Golgi area also re-

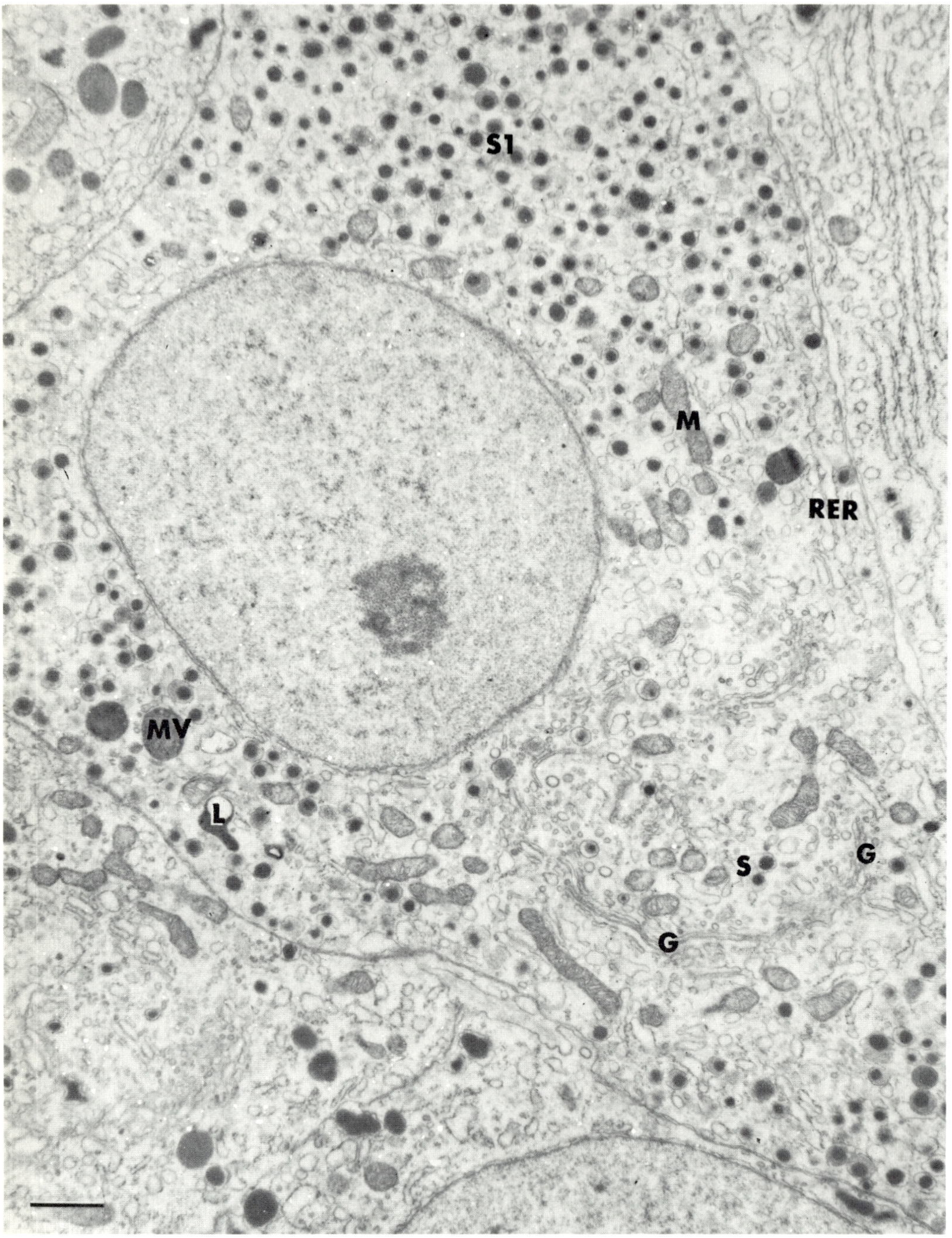

Fig. 4.6. An electron micrograph of a section through the pituitary gland of a female rat showing a stimulated LH cell. Extensive rough endoplasmic reticulum (RER) is shown throughout this hypertrophied cell. A lysosome (L), multivesicular body (MV), mitochondria (M) with dense cristae, and several Golgi complex areas (G) with granules (S) in various stages of formation are shown. Irregularly shaped mature granules (S1) are also found. × 11,000. (Veronal acetate buffered osmic acid)

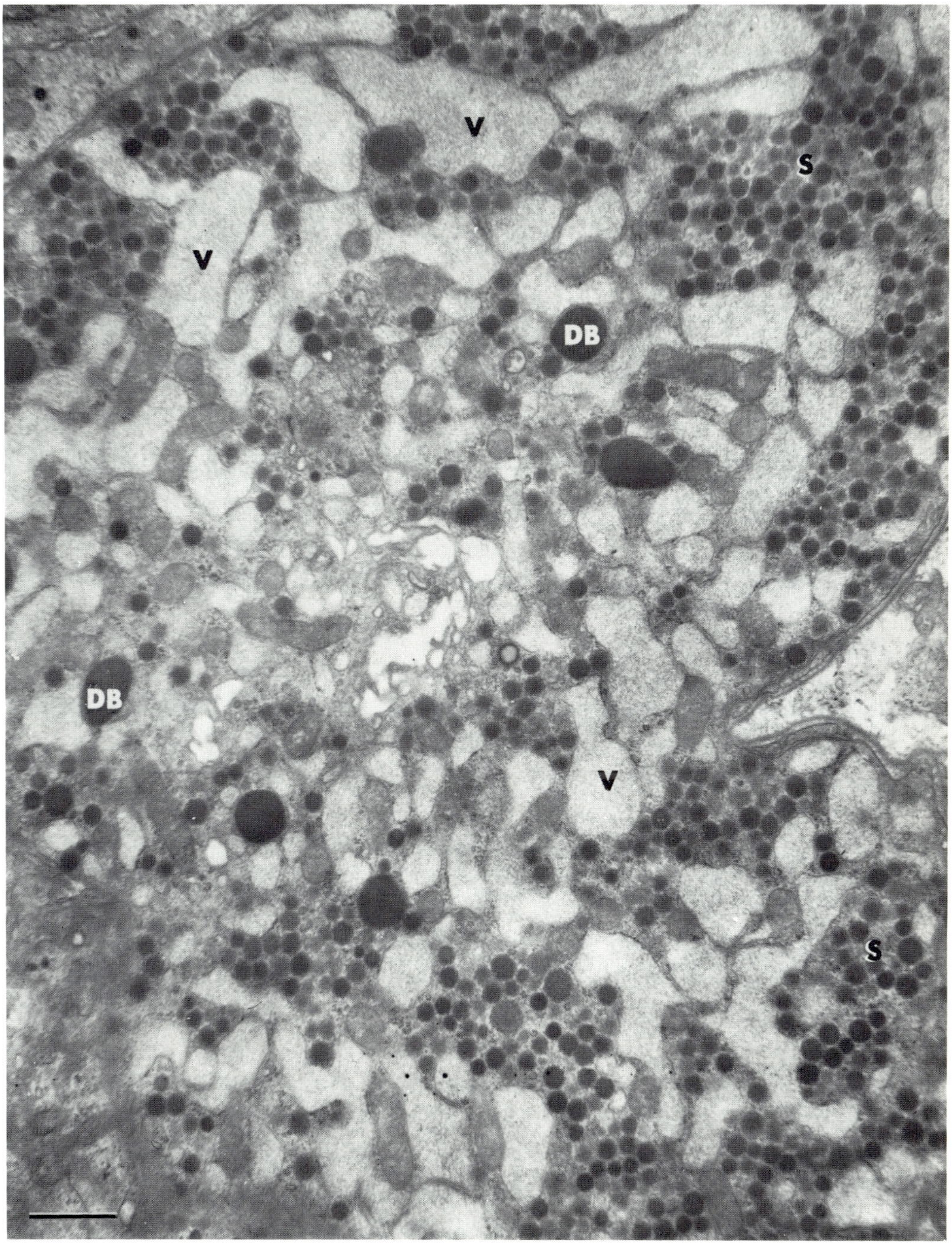

Fig. 4.7. A follicle-stimulating hormone cell in a male rat pituitary after 7 days of castration. There is an abundance of granules (S), several dense bodies (DB), and many vacuoles (V). The large opaque bodies are not present. $\times$ 12,000. (Veronal acetate buffered osmic acid)

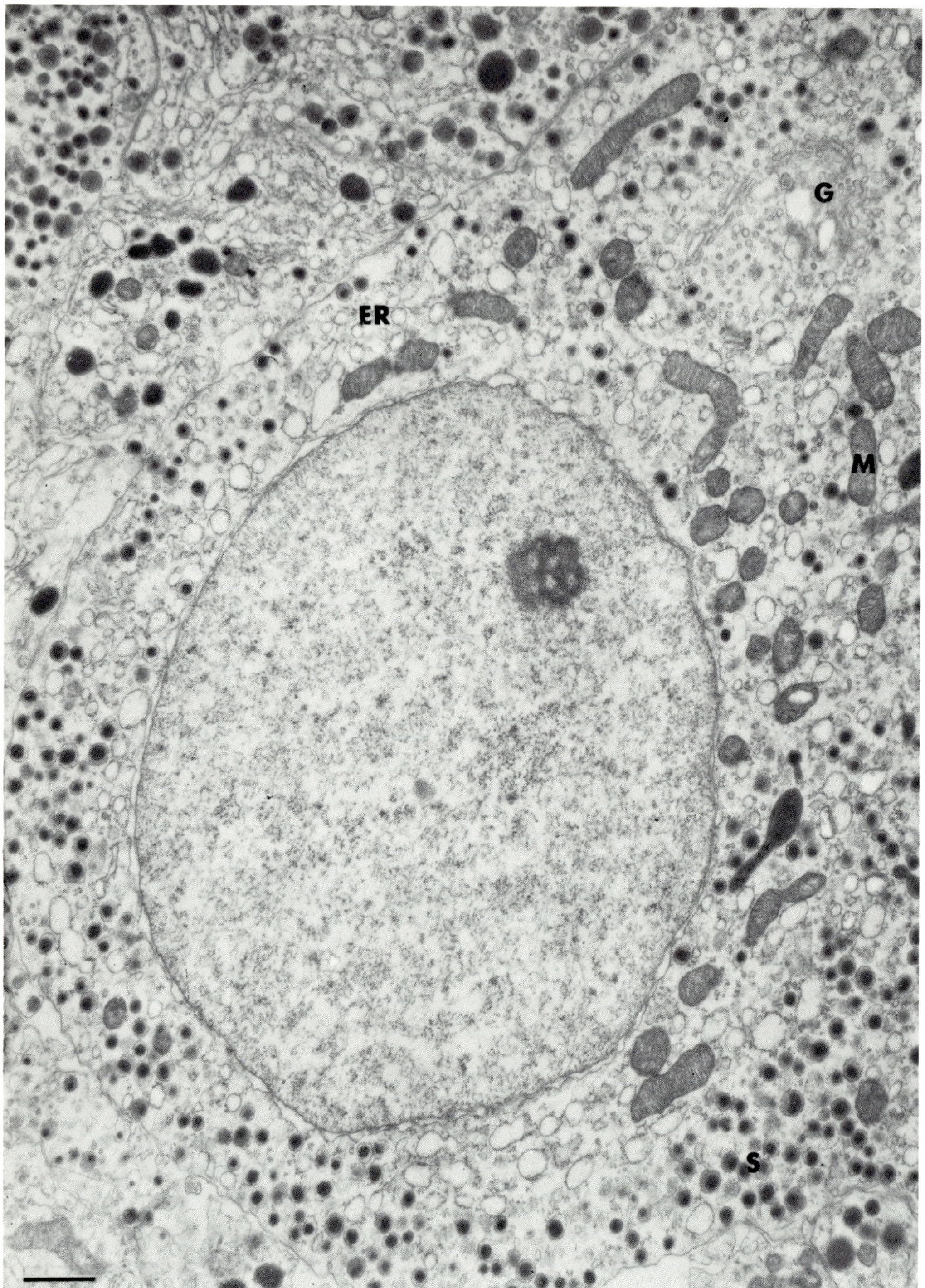

Fig. 4.8. A luteinizing hormone gonadotrope after 7 days of castration. There is an increase in granulation (S), a Golgi complex (G), mitochondria with dense cristae (M), and several dilated vesicles of endoplasmic reticulum (ER). × 11,000. (Veronal acetate buffered osmic acid)

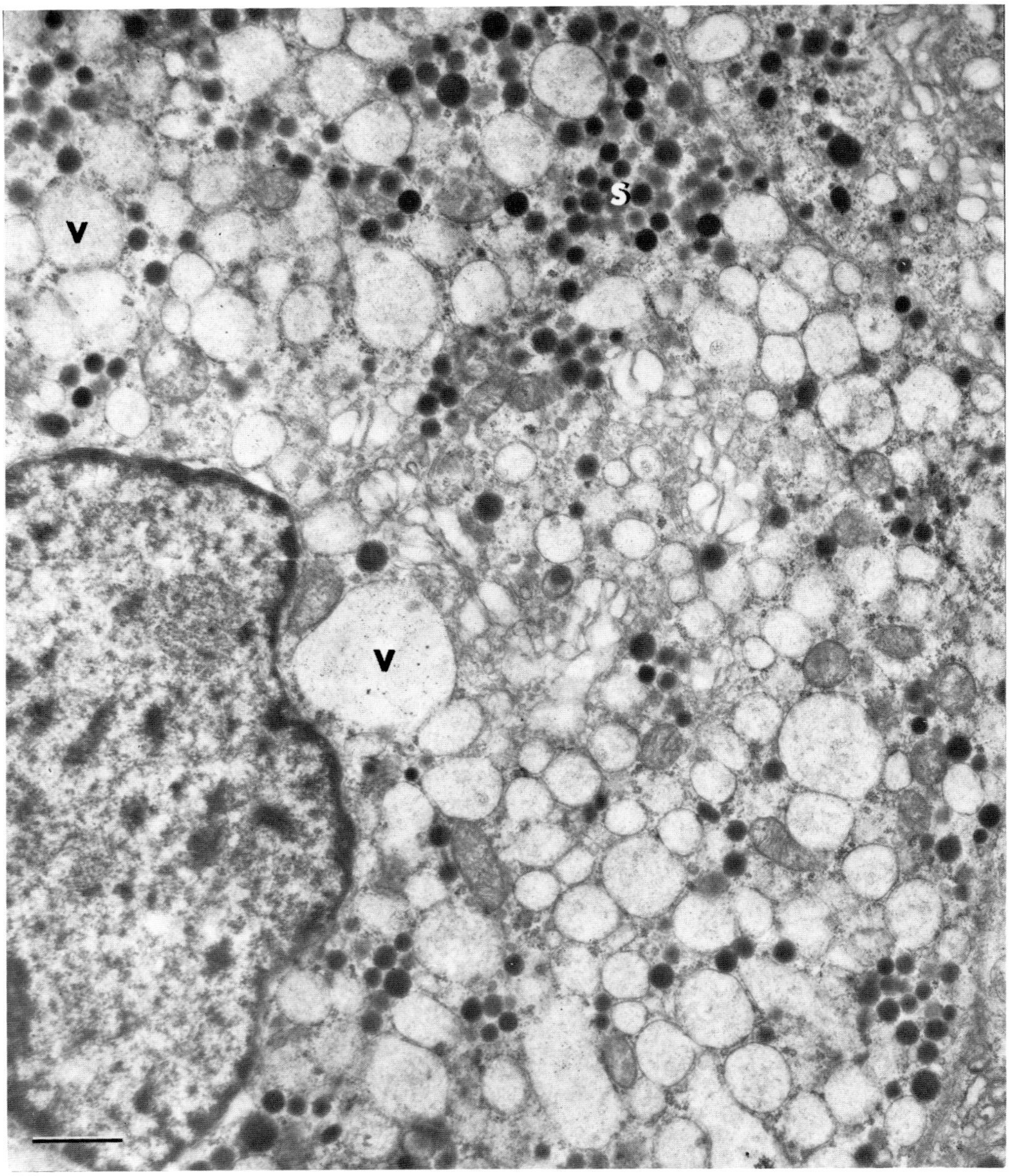

Fig. 4.9. A FSH gonadotrope 14 days after castration in the male rat pituitary. This cell is large, contains granules (S), but shows an increase in vacuoles (V). × 12,500. (Veronal acetate buffered osmic acid)

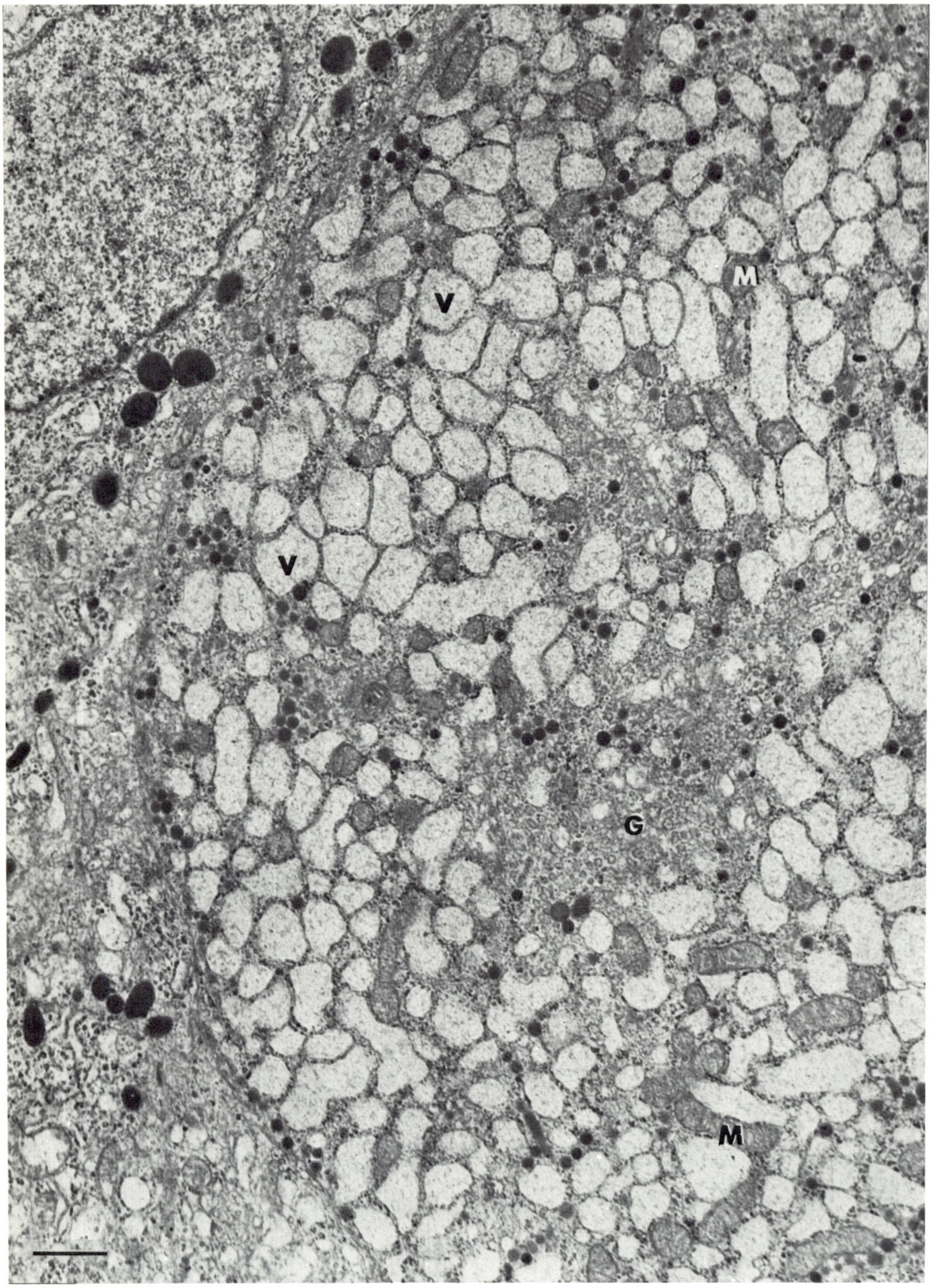

Fig. 4.10. A FSH cell 60 days after castration. This cell is greatly enlarged but vacuolation (V) has increased. A Golgi complex (G) is shown in the center of the cell. Mitochondria (M) have dense cristae. × 10,500. (Veronal acetate buffered osmic acid)

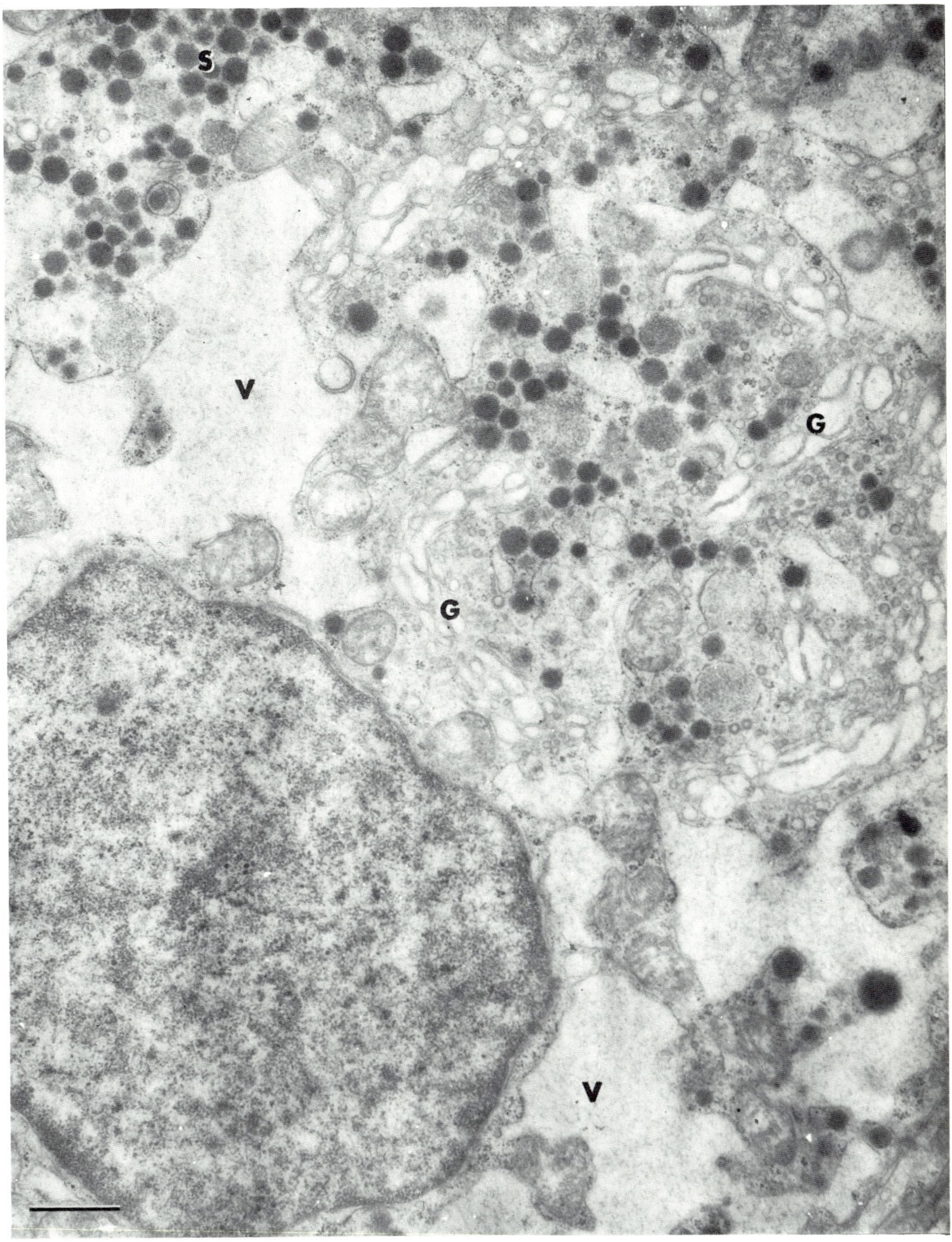

Fig. 4.11. A part of an enlarged LH cell in the male rat pituitary gland is shown 14 days after castration. An increase in granules (S), dilated Golgi complexes (G), and great vacuolated areas (V) are present. × 13,400. (Veronal acetate buffered osmic acid)

mained distinct up to 30 days following castration but then became more inconspicuous. The endoplasmic reticulum of these LH cells hypertrophied and became more vacuolated. The vesicles of the endoplasmic reticulum coalesced to form vacuoles, and 30–60 days after castration there were numerous LH cells with one large vacuole or "lake" in the middle of the cell (Fig. 4.12). Such cells are called "signet ring" cells which can be further characterized by having the nucleus and cytoplasm oriented to one side near the cell membrane. This large vacuole often contained some colloidlike substance. Mitochondria by this stage of castration were somewhat hypertrophied with the round type predominating. These observations indicated that LH cells undergo more profound changes in ultrastructure after castration than FSH cells.

The Ultrastructure of the Female Rat Anterior Pituitary Gland at Different Stages of the Estrous Cycle

In describing changes of gonadotropes during the estrous cycle two major criteria will be used: *(1)* number of secretory granules in the cells, and *(2)* development of endoplasmic reticulum and Golgi complex. In the mature female rat there were more LH than FSH gonadotropes and, although granulation or hormone content may change during the estrous cycle, cell numbers were not observed to change.

At middiestrus, FSH gonadotropes were somewhat smaller, poorly granulated, exhibited poorly developed endoplasmic reticulum, and had fewer attached ribosomes. At proestrus the cells became more granulated and the rough endoplasmic reticulum and active Golgi complexes became more prominent. At midproestrus these FSH cells were filled with granules and large bodies (Fig. 4.13). Golgi complexes and endoplasmic reticulum were rather well developed (Fig. 4.14). During late proestrus and early estrus the cells were degranulated and appeared vacuolated and chromophobelike (Fig. 4.15). At metestrus the FSH cells were degranulated, but by early diestrus (Fig. 4.16) they were again well granulated.

LH cells were well granulated during diestrus. The endoplasmic reticulum and Golgi complexes appeared to be active; they were often occluded (Fig. 4.17) by the great abundance of secretory granules at this time. They began to degranulate during midproestrus and were fully degranulated (Fig. 4.18) by the critical period, estrus. In metestrus (Fig. 4.19) the endoplasmic reticulum and Golgi complexes again appeared active and by diestrus this cell was well granulated. It would appear, therefore, that both gonadotropes degranulate during the critical period.

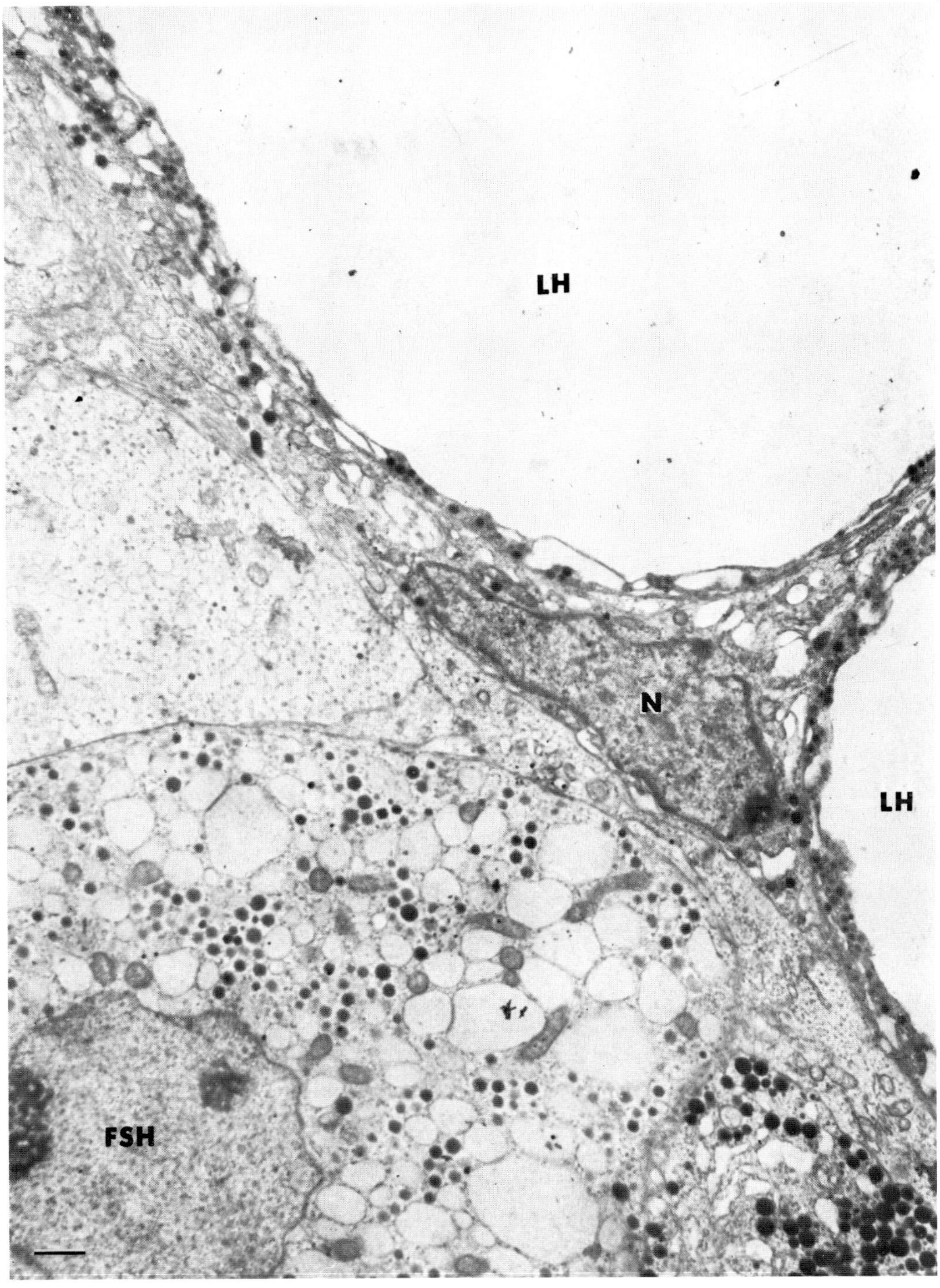

Fig. 4.12. Two LH signet ring cells are shown in this electron micrograph of a male rat pituitary after 60 days of castration. A FSH cell with vacuolar endoplasmic reticulum is also shown. The LH cells are greatly hypertrophied and the cytoplasm and nucleus (N) are displaced toward the cell membrane. The signet ring or lake often is filled with colloid material. × 8250. (Veronal acetate buffered osmic acid)

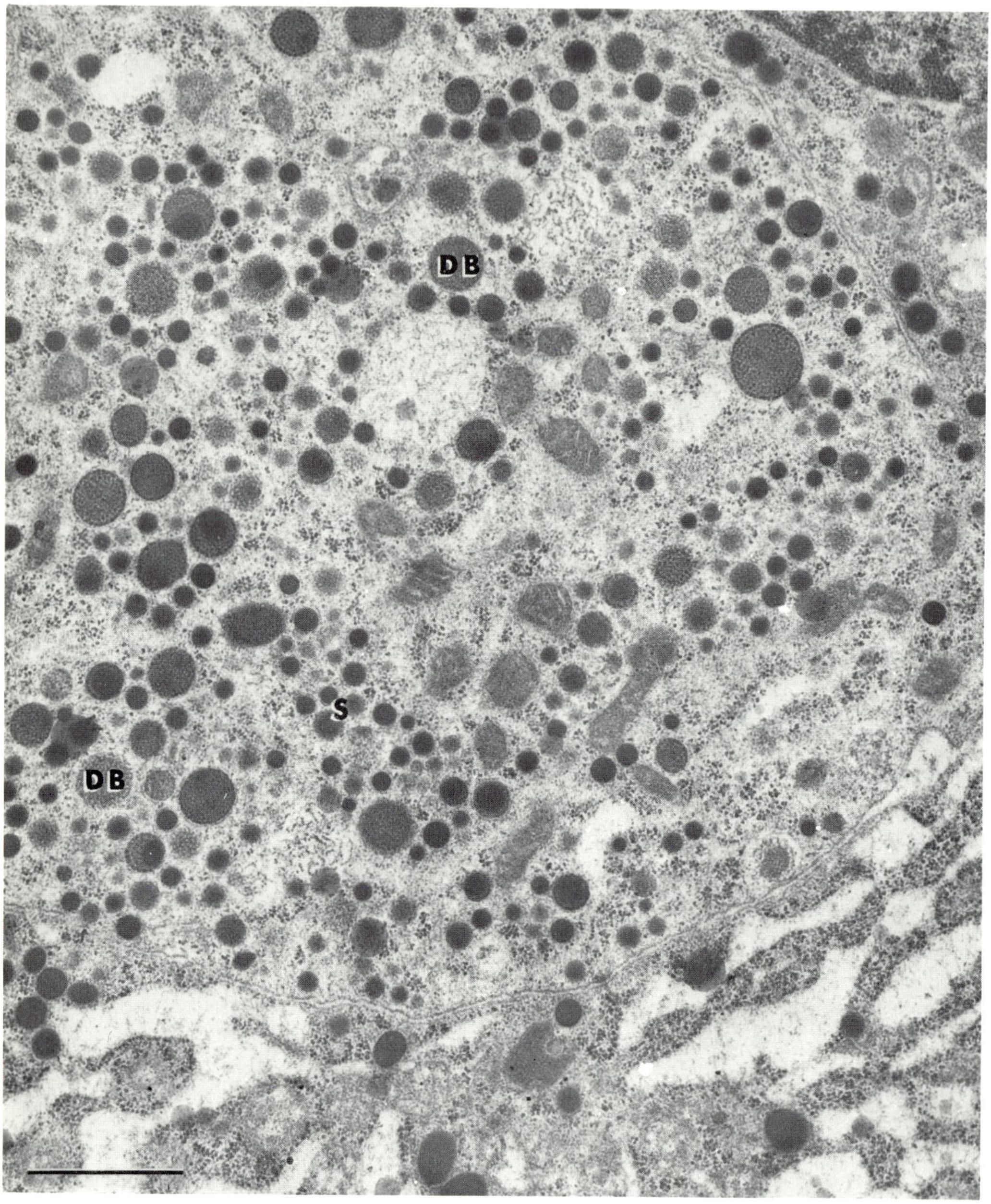

Fig. 4.13. A FSH gonadotrope at midproestrus. This cell contains an abundance of granules (S) and large bodies (DB) of varying electron density: × 22,000. (Glutaraldehyde–paraformaldehyde)

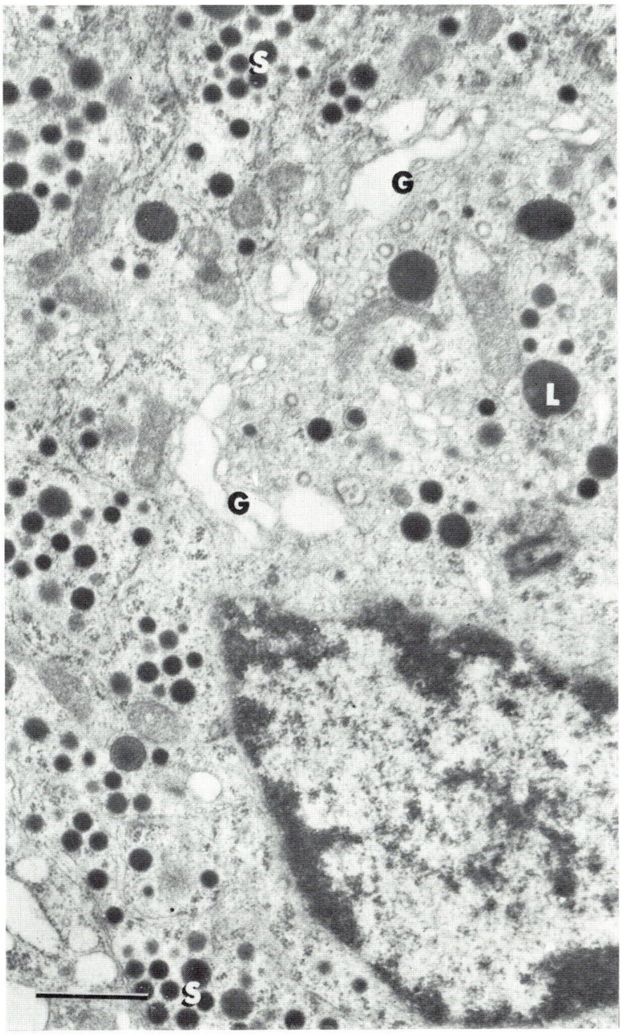

Fig. 4.14. An enlarged FSH cell at proestrus. The Golgi complexes (G) are dilated. The opaque light staining bodies appear to be dark lysosomes (L). At the poles of the cell granules (S) are prominent. × 15,000. (Glutraldehyde–paraformaldehyde)

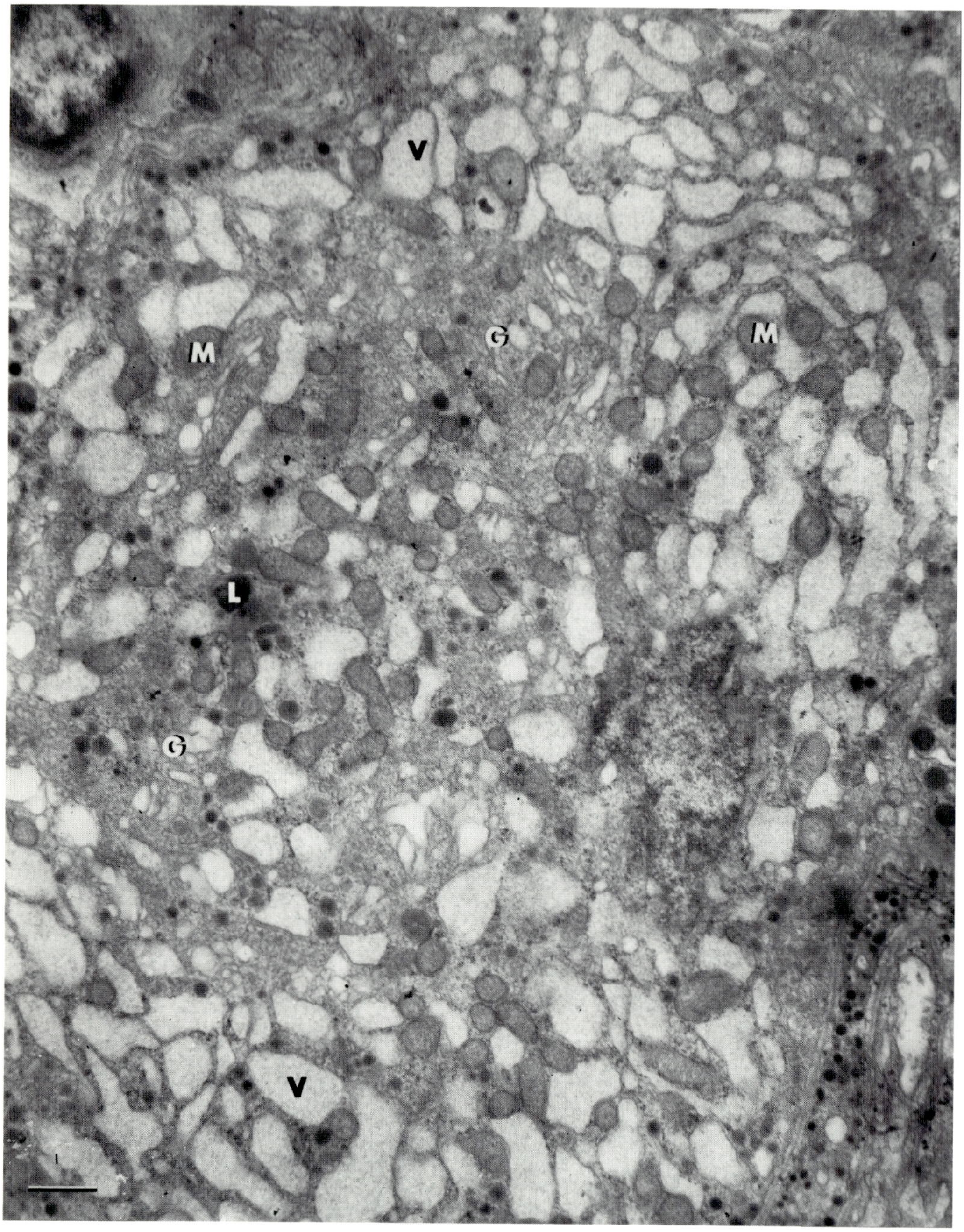

Fig. 4.15. A FSH cell during early estrus. The cell is enlarged with many vacuoles (V), a preponderance of round mitochondria (M), a lysosome (L), and several Golgi complexes (G). $\times$ 10,000. (Glutaraldehyde–paraformaldehyde)

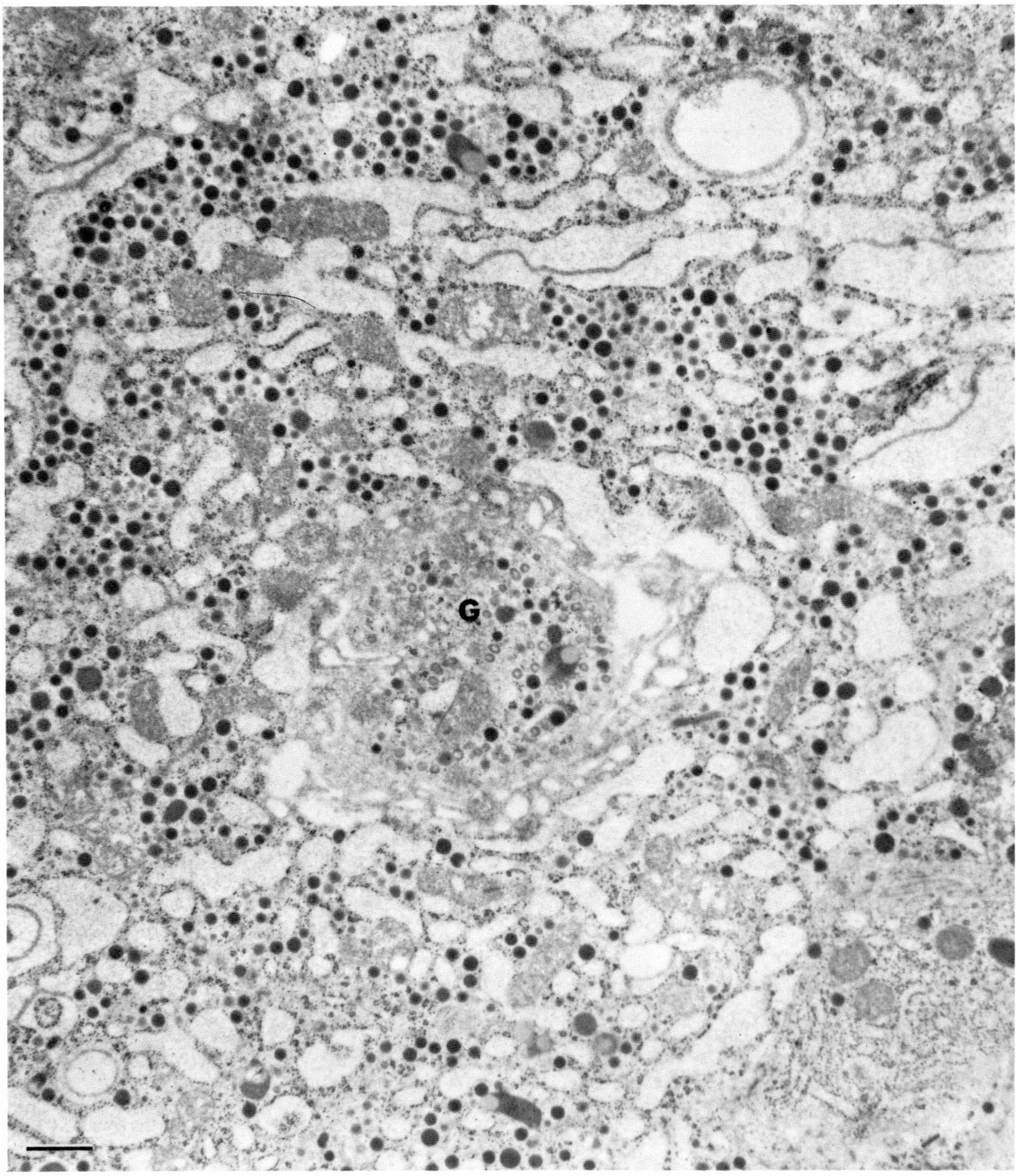

Fig. 4.16. A FSH gonadotrope at early diestrus. This enlarged cell has an abundance of granules and an active Golgi complex (G). × 8800. (Glutaraldehyde–Paraformaldehyde)

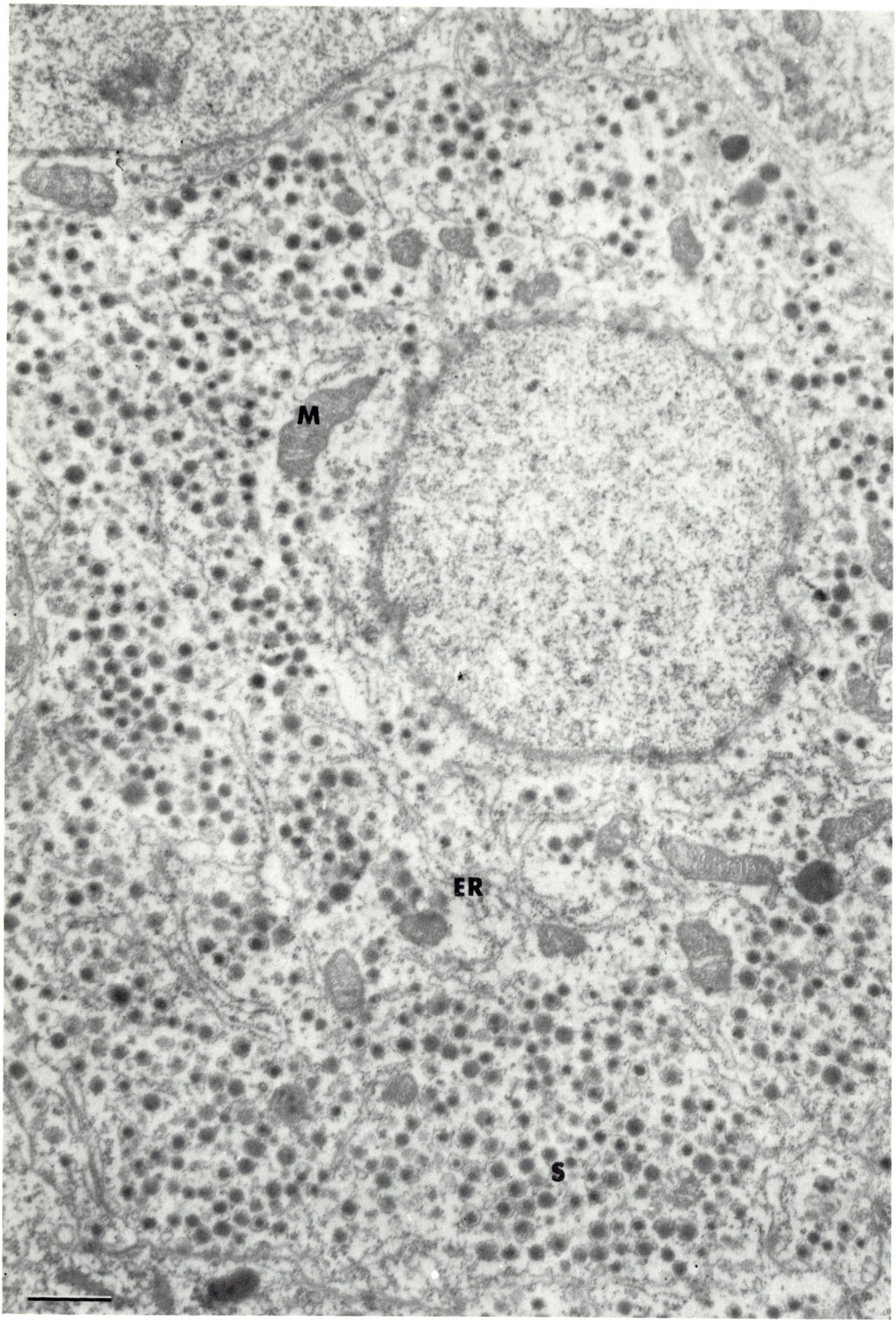

Fig. 4.17. A LH gonodatrope at late diestrus. This cell is enlarged, contains an abundance of granules (S), mitochondria with dense cristae (M), and several cisternae of endoplasmic reticulum (ER). × 13,500. (Glutaraldehyde–paraformaldehyde)

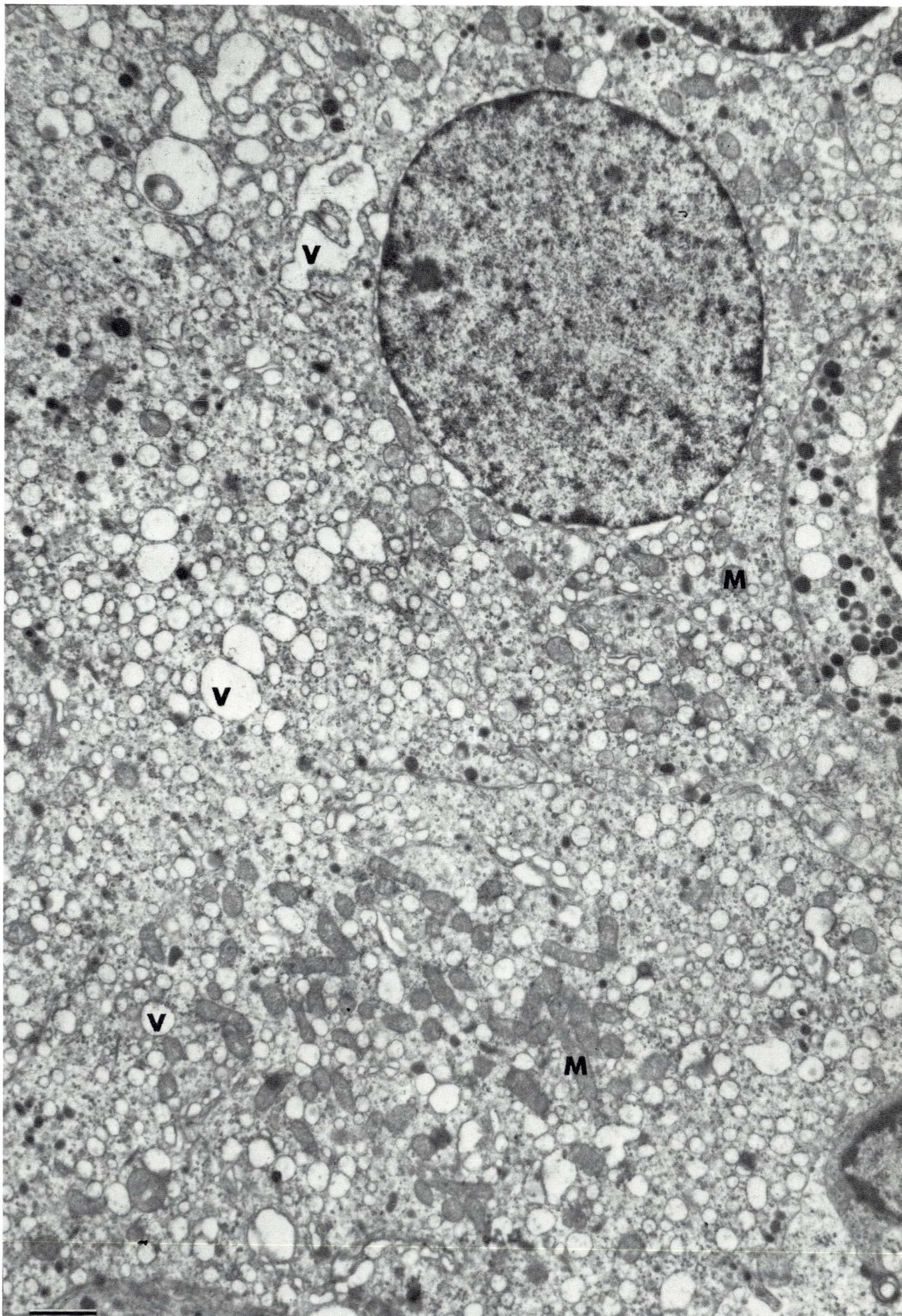

Fig. 4.18. At estrus the LH gonadotrope is degranulated; its cytoplasm is vacuolated (V) but mitochondria (M) have dense cristae. × 10,000. (Glutaraldehyde–paraformaldehyde)

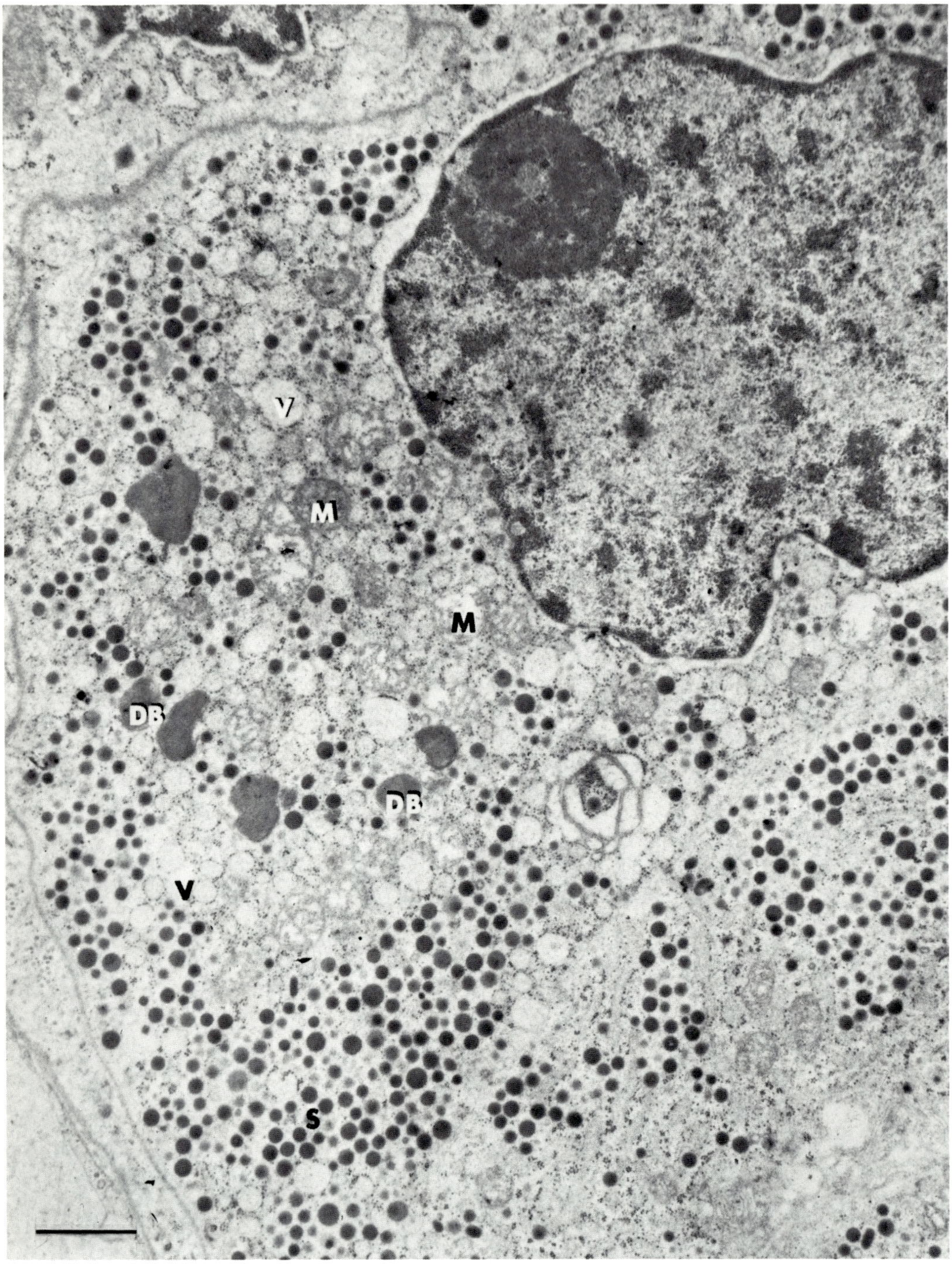

Fig. 4.19. The LH cell at metestrus has begun to regranulate. Granules (S) Predominate at the lower pole of the cell. Vacuolated (V) areas still are evident. Mitochondria (M) with discontinuous cristae and dense bodies (DB) are shown. × 15,000. (Glutaraldehyde–paraformaldehyde)

Discussion of Estrous Cycle

Hunt and Hunt (1966) reported in their autoradiographic studies that the greatest labeling with ^{3}H-thymidine took place at late estrus and 12–24 hours after ovulation. This is the time at which the great turnover of pituitary proteins occurs. They postulated that there is an increase in mitotic activity at this time or an increase in synthesis of messenger RNA and perhaps other RNA species, resulting in the increase of protein synthesis. Since no mitotic figures were seen microscopically, it appears the latter postulation may be more accurate.

Roos (1969) found that atropine or meprobamate prevented the secretion of both FSH and LH. Both types of gonadotropes were well granulated. This indicates that these two hormones act synergistically during ovulation.

Caligaris *et al.* (1967) reported that FSH concentrations in the pituitary were highest on the morning of proestrus, before 9 AM, and declined sharply by 5 PM just after the critical period and then increased gradually until the next proestrus. Ovariectomy before 9 AM, pentobarbitol, or atropine prevented the decrease in pituitary FSH. Progesterone administered to ovariectomized rats caused a release of FSH and a coincident drop in pituitary FSH. Estradiol had no effect in this regard. These workers believe that a neurogenic mechanism possibly triggered by progesterone is involved in causing the decrease or release of FSH at late proestrus.

Monroe *et al.* (1969) observed that serum LH was increased 50 times on the afternoon of proestrus and the basophils were degranulated. The LH in the blood serum started to increase as early as 1 PM and was still elevated at 8 PM of proestrus. The highest concentration of LH was between 3:00 and 6:00 PM.

Ultrastructure of Gonadotropes after Adrenalectomy and Castration and Other Treatments

The cytological appearance of the pituitary after double extirpation was quite similar to adrenalectomy or castration alone (Fig. 5.18). Changes in FSH and LH cells were not as pronounced as after castration alone. Fewer signet ring cells were observed. Rokhlina (1940) reported that castration changes did not occur in most gonadotropes until 30 days after double extirpation, which was 14 days later than the time changes occurred following castration alone.

After administration of adrenal blocking agents, gonadotropes were generally normal but smaller. After α-ethyltryptamine these cells were well granulated and were larger than normal.

DISCUSSION OF GONADOTROPES

The identification of the two gonadotropes has been controversial. Kragt and Ganong (1967), Gay and Bogdanove (1967), Parlow (1968), and others have reported that there is as much as 20 times more FSH in the glands of adult male rats than in the adult female rats. These workers also reported that the LH content in the adult males is usually lower than that in the adult female rats. We observed more FSH cells and less LH cells in adult male rats than in females, which correlates well with the results of these workers. Barnes (1962b) and Herlant (1964) described changes in pituitary cells during the estrous cycles of mice and moles respectively and reported that FSH cells degranulate before LH cells.

In our study the cell type that was abundant in the male was tentatively identified as the FSH gonadotrope. This identification was also based on physiological and bioassay data. It was found that the FSH cell was large, round, and was the second most abundant pituitary cell type in the male but it was more difficult to identify in the female pituitary gland. The FSH cells which were abundant throughout the medial part of the male pituitary contained small granules of 200 mμ maximum diameter, and also light-staining, amorphous bodies that were about 1.0 μ in diameter. These large, light-staining bodies were not always present; if they were not, there were large vacuoles in the endoplasmic reticulum where they originally had been. These large, light-staining bodies were only found in the FSH cells and appeared darker after glutaraldehyde fixation.

These light bodies were not found in glands of castrate rats, but more electron dense bodies were present which appeared to have the characteristics of lysosomes. It is postulated that these lighter bodies may fuse with other vesicles and be transformed into lysosomes. Cardell (1961) observed similar large bodies in the salamander pituitary and reported that these bodies disappeared at the breeding season, which was coincident with the degranulation of gonadotropes. He concluded that these bodies might be secretory granules or lysosomes.

Herlant (1965) reported that the light bodies are glycoprotein in nature and stain with PAS. Nakayama, *et al.* (1970) have observed these light staining bodies in the Golgi zone. They postulate that these structures may be formed in the Golgi apparatus or by the coalescence of small secretory granules. Acid phosphatase was not found to be associated with these bodies. These two groups of workers suggest that hormones may be stored in these organelles. The bioassay of the large granule fraction presented here and reported by others (Perdue and McShan, 1962;

Hymer and McShan, 1963) have not shown any gonadotropic hormone activity. It appears unlikely that these organelles are secretory granules.

In the present study these large, amorphous bodies were observed in the Golgi complex in various stages of formation. In this connection Novikoff *et al.* (1964) have shown that lysosomal hydrolases were packaged into lysosomes in the Golgi complex. Wächtler and Pearse (1966) have located by histochemical means five lysosomal enzymes in FSH cells of the amphibian pituitary that were not found in the other cell types. They postulated that lysosomal hydrolases may play a role in the transformation of precursor FSH to the functionally active hormone. This is an attractive postulation since Dasgupta *et al.* (1967) have shown that in many mammalian species 80% of the total corticotropic activity is in the form of procorticotropin, a precursor of corticotropin. A similar situation may exist for FSH in that a pre-FSH may be transformed to FSH. This is a possible explanation for the function of these large, light-staining bodies, but more evidence is necessary to identify the role they play in the synthesis or release of FSH.

Gonadotropes After Castration

It was observed that FSH cells did not undergo changes due to castration. These findings are in agreement with those of Parlow (1968) that LH content of the pituitary and not FSH increased following castration in male rats. Ramirez and McCann (1963) found that LH production and release predominate after castration in male rats. Bogdanove *et al.* (1964) also reported that the FSH concentration of the male rat pituitary was not changed 49 days after castration, but LH was increased eightfold.

Although FSH cells were abundant in the male gland, LH cells were difficult to find. They were more abundant in the medial anterior region of the gland, while FSH cells were found quite extensively in medial and to some extent in lateral regions. This agrees with the work of Smelser (1944), Purves and Griesbach (1955), and Kraicer and Logothetopoulos (1961), who have shown the LH gonadotrope to be in the medial region.

One week after castration the LH cell type began to hypertrophy. After 14 days this cell enlarged more dramatically with the fusion and hypertrophy of vesicles and/or vacuoles of the endoplasmic reticulum. According to Herlant (1964) the resulting large vacuoles that form are often filled with a PAS staining material. By 1 month after castration the majority of LH cells were transformed into signet ring castration cells with large vacuoles or lakes in the middle of them. Often these were filled with colloidlike material from which LH may be stored and released. More dramatic changes in LH cells have been reported after long-term castration experiments by Foncin and LeBeau (1966) in the human, Dekker (1967) in the hamster, and Barnes (1963) in the mouse. It appears from these castration studies that an FSH cell has been identified and that it is different from the LH cell.

Recently, further evidence for the identification of LH and FSH cells has been reported by Kurosumi and Oota (1968). Their identification of gonadotropes is similar to that given here. These workers produced persistent estrus and diestrus in rats by administration of estrone at birth for 5 and 30 days, respectively. After 5 months the ovaries in persistent estrus rats had well developed Graafian follicles but no corpora lutea. The pituitary of these animals had slightly granulated FSH cells but the LH cells were well granulated. The release of these LH granules during persistent estrus was presumably suppressed. In the persistent diestrus rat, the FSH cells were rather atrophic but LH cells were like the controls. Although there were follicles in the ovaries of these animals, well-developed interstitial tissue predominated. It would appear that the original identification of gonadotropes by Barnes (1963) was correct and indicates that the cells producing FSH and LH are distinct cell types.

FSH AND LH ACTIVITY OF ISOLATED SECRETORY GRANULE FRACTIONS

In the past, ^{32}P uptake in chick testes or increase in ovarian weight were the procedures used to assay for gonadotropic hormones (Hartley *et al.*, 1960; Hymer and McShan, 1963). These procedures did not distinguish between FSH and LH activities. Because of this, the Parlow (1961) ascorbic acid depletion assay for LH and the Steelman–Pohley assay (1953) for FSH were used for the assay of these hormones in the granule fractions.

Follicle-Stimulating Hormone

The FD2HSP pellet (Fig. 4.20) contained granules with FSH and LH activities. The recovery of FSH from the original FS1 placed on the gradient was 82.7% and of this 55.7% was recovered in the FD2HSP pellet. Low levels of FSH activity were also present in the SA soluble zone, in the FD2LSP pellet, and the D1HSP pellet. FSH was not found in the large acidophilic granule fraction.

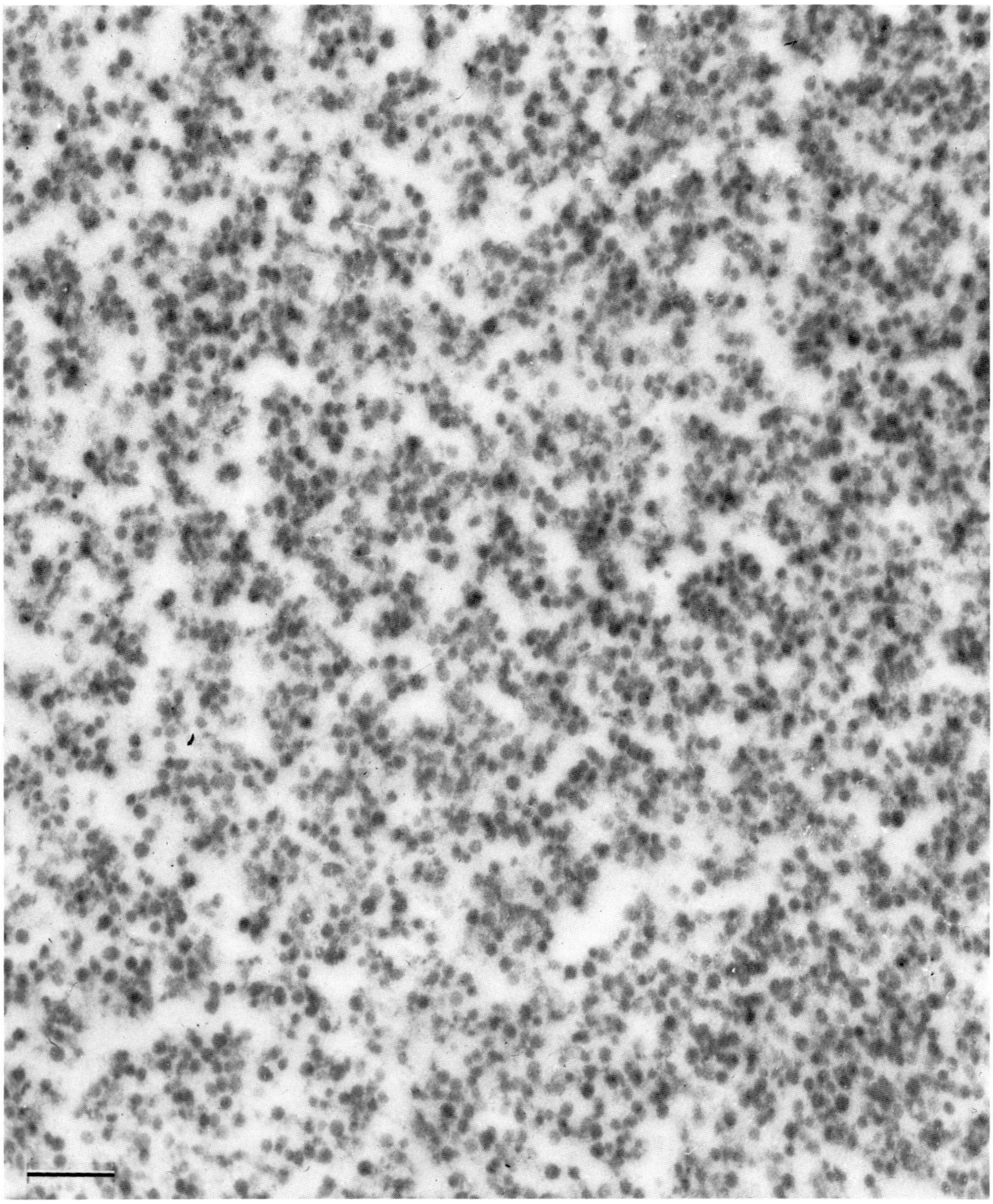

Fig. 4.20. A section through the FD2HSP pellet is shown in this micrograph. It consists of granules that range in size from 75 to 218 mμ. $\times$ 12,000. (Veronal acetate buffered osmic acid)

Luteinizing Hormone

Zones from three different granule fractionations were assayed for LH activity. Fifty-one percent of the LH recovered from the gradient was found in the FD2HSP pellet (Fig. 4.20). Over 84% of the original LH activity of the FS1 fraction applied to the gradient was recovered. The soluble SA zone contained 23.1% of the remaining LH activity, but only 7.7% was associated with the FD2LSP pellet.

Diameters of Sections of Granules

Based on 500 measurements, granules from follicle-stimulating hormone cells had a mean diameter of 133. mμ The LH cell contained granules with a mean of 145. mμ The FD2HSP pellet with which these hormones were associated had a mean of 150 mμ. These measurements are similar to those recorded by others as reviewed by McShan and Hartley (1965).

Summary

Electron microscopic observations of anterior pituitary cells before and after castration have helped to identify the FSH and LH gonadotropes. After castration LH gonadotropes of male rat pituitaries are greatly enlarged and ultimately form signet ring cells. FSH cells do not change appreciably during early castration and tend to degranulate after long-term castration. The initial FSH and LH cell identification by Barnes (1962a, 1963) has been further substantiated by data obtained from persistent estrus and diestrus rats by Kurosumi and Oota (1968), from injection of methylandrostenediol experiments by Nakayama *et al.* (1970), and by our data. There still is a question concerning the identity of the large light staining bodies in the FSH cells. Considerable evidence supports the concept that these bodies may be lysosomes. However, they may contain pre-FSH or FSH hormone.

Nakane (1970) in his immunochemical studies at the electron microscopic level has indicated that FSH and LH may be produced by one and not two cells. His contrary findings may be due to the fact that the FSH and LH he used were not highly purified and that the antibodies formed were not specific to one of these gonadotropes but rather reacted with both FSH and LH cells.

Added evidence that there are two distinct gonadotropes is given by the work of Shiino *et al.* (1972). These workers administered LH releasing factor (LRF) to persistent estrus rats. Not only did they observe an increase in serum LH, and not FSH, after the LRF was given but they also showed electron micrographs of the release of granules from LH cells.

REFERENCES

Addison, W. H. F. (1917). The cell changes in the hypophysis of the albino rat after castration. *J. Comp. Neurol.* **28**, 441.

Bain, J., and Ezrin, C. (1970). Immunofluorescent localization of the LH cell of human adenohypophysis. *J. Clin. Endocrinol. Metab.* **30**, 181.

Barnes, B. G. (1962a). Electron microscopic studies on the cytology of the mouse anterior pituitary. *Endocrinology* **71**, 618.

Barnes, B. G. (1962b). The fine structure of the mouse anteriör pituitary during the estrous cycle. *Electron Microsc., Proc. Int. Congr., 5th, 1962* Vol. 2, p. 7710.

Barnes, B. G. (1963). The fine structure of the mouse adenohypophysis in various physiological states. In "Cytologie de l'Adenohypophyse" (J. Benoit and C. Da Lage, eds.) , p. 91. Editions du Centre National de La Recherche Scientifique, Paris.

Barnett, R. H., Ladman, A. J., McAllaster, N. J., and Siperstein, E. R. (1956). The localization of glycoprotein hormones in the anterior pituitary glands of rats investigated by differential protein solubilities, histological stains and bioassay *Endocrinology* **59**, 398.

Bogdanove, E. M., Parlow, A. F., Bogdanove, J. N., Bhargava, I., and Crabill, E. V. (1964). Specific LH and FSH bioassays in rats with hypothalamic lesions and accessory sex gland hypertrophy. *Endocrinology* **74**, 114.

Caligaris, L., Astrada, J. J., and Taleisnik, S. (1967). Pituitary FSH concentration in the rat during the estrous cycle. *Endocrinology* **81**, 1261.

Cardell, R. R. (1961) . Observations of ultrastructure of the salamander pituitary gland. *ASB Bull.* **8**, 27.

Catchpole, H. R. (1949). Distribution of glycoprotein hormones in the anterior pituitary gland of the rat. *J. Endocrinol.* **6**, 218.

Chow, B. F. (1943). The chemistry of thylakentrin, the follicle stimulation hormone of the anterior pituitary. *Ann, N. Y. Acad. Sci.* **43**, 309.

Chowdhury, M., Steinberger, A., and Steinberger, E. (1971). Correlation between LH and FSH production and staining characteristics of rat pituitary cells grown in culture. *Anat. Rec.* **169**, 296.

Comte, L. (1898). Contribution à l'étude de l'hypophyse humaine. Ph. D. Thesis, Lausanne.

Dasgupta P. R., Margalis, S. A., and Dorfman, R. I. (1967). On the presence of precorticotropin—A precursor of corticotrophin in various mammalian hypophyses. *Acta Endocrinol.* **55**, 31, *(Copenhagen)*

Dekker, A. (1967). Pituitary basophils of the Syrian hamster: An electron microscopic investigation. *Anat. Rec.* **158**, 351.

Della Corte, F., and Biondi, A. (1964). *Riv. Biol.* **57**, 369.

Ducommun, S. (1965). Etude autoradiographique de l'incorporation de la proline tritise par les cellules basophiles de l'adenohypophyse du rat. *Ann. Endocrinol.* **26**, 385.

Engle, E. T. (1929). The effect of daily transplants of the anterior lobe from gonadectomized rats on immature test animals. *Amer. J. Physiol.* **88**, 101.

Evans, H. M., and Simpson, M. E. (1929). A comparison of anterior hypophyseal implants from normal gonadectomized animals with reference to their capacity to stimulate the immature ovary. *Amer. J. Physiol.* **89**, 371.

Farquhar, M. G. (1955). Further evidence for the existence of two types of gonadotropes in the pituitary of the rat. *Anat. Rec.* **121**, 394.

Farquhar, M. G., and Rinehart, J. F. (1954). Electron microscopic studies of the anterior pituitary gland of castrate rats. *Endocrinology* **54,** 516.

Fevold, H. L. (1943). The luteinizing hormone of the anterior lobe of the pituitary gland *Ann. N. Y. Acad. Sci.* **43,** 321.

Fevold, H. L., Hisaw, F. L., and Leonard, S. L. (1931). The gonad stimulating and the luteinizing hormones of the anterior lobe of the hypophysis. *Amer. J. Physiol.* **97,** 291.

Fevold, H. L., Hisaw, F. L., Hellbaum, A., and Hertz, R. (1933). Sex hormones of the anterior lobe of the hypophysis. *Amer. J. Physiol.* **104,** 710.

Fichera, G. (1905). Sur l'hypertrophie de la glande pituitaire consecutive à la castration. *Arch. Ital. Biol.* **43,** 405.

Fogel, M., and Koffler, D. (1964). Immunofluorscent localization of LH and FSH of fresh, frozen and formalin fixed sections of the human anterior pituitary. *Fed. Proc., Fed. Amer. Soc. Exp. Biol.* **23,** 411.

Foncin, J., and LeBeau, J. (1966). Cellule de castration et cellules FSH dans l'hypophyse humaine vues au microscope electronique. *J. Microsc. (Paris)* **5,** 523.

Fraenkel-Conrat, H., Li, C. H., and Simpson, M. E. (1943). Pituitary Gonadotropins. In "Essays in Biology," p. 183. Univ. of California Press, Berkeley.

Gay, V. L., and Bogdanove, E. M. (1967). Observations on the clearance of endogenous LH from the plasma of previously castrated rats. *Fed. Proc., Fed. Amer. Soc. Exp. Biol.* **26,** 533 (abstr.).

Girod, C., and Dubois, P. (1965). Etude ultrastructurale des cellules gonadotropes antéhypophysaires chez le hamster. *J. Ultrastruct. Res.* **13,** 212.

Giroud, A., and Martinet, M. (1948). Localisations de certaines fonctions dans le lobe antérieur de l'hypophyse. *Ann. Endrocrinol.* **9,** 343.

Greep, R. O., and Jones, I. C., (1950). Steroid control of the pituitary function. *Recent Progr. Horm. Res.* **5,** 197.

Greep, R. O., van Dyke, H. G., and Chow, B. F., (1941). Some biological properties of metakentrin and thylakentrin *Amer. J. Physiol.* **133,** 303.

Guyer, M. F., and Claus, E. G. (1937) . Vacuolation of the anterior pituitary gland following castration, implantation of cancer tissue and thyroidectomy. *Anat. Rec.* **67,** 145.

Halmi, N. S. (1950). Two types of basophils in the anterior pituitary of the rat and their respective cytophysiological significance. *Endocrinology* **47,** 289.

Halmi, N. S. (1952). Two types of basophils in the rat pituitary: Thyrotropes and gonadotropes vs. beta and delta cells. *Endocrinology* **50,** 140.

Hartley, M. W. (1959). Purification, isolation and description of cytoplasmic granules associated with gonadotropic hormones of the anterior pituitary gland of the rat. Ph.D. Thesis, University of Wisconsin, Madison.

Hartley, M. W., McShan, W. H., and Ris, H. (1960). Isolation of cytoplasmic pituitary granules with gonadotropic activity. *J. Biophys. Biochem. Cytol.* **7,** 209.

Hellbaum, A. A., McArthur, L. G., Campbell, P. J., and Finerty, J. C. (1961). The physiological fractionation of pituitary gonadotropic factors correlated with cytological changes. *Endocrinology* **68,** 144.

Herlant, M. (1943). Recherches sur la localisation histologique des hormones gonadotropes femelles au niveau de l'hypophyse antérieure. *Arch. Biol.* **54,** 225.

Herlant, M. (1960). Etude critique de deux techniques nouvelles destinées à mettre en évidence les différentes catégories cellulaires présentes dans la glande pituitaire. *Bull. Microsc. Appl.* **10,** 37.

Herlant, M. (1964). The cells of the adenohypophysis and their functional significance. *Int. Rev. Cytol.* **17**, 299.

Herlant, M. (1965). The present state of knowledge concerning the cytology of the anterior lobe of the hypophysis. *Proc. Int. Congr. Endocrinol., 2nd, 1964* p. 468.

Herlant, M., and Ectors, F. (1970). Identification des celles LH du porc en microscopie optie, microscope électronique et par immunofluorescence. *C. R. Acad. Sci.* **269**, 368.

Hildebrand, J. E., Rennels, E. G., and Finerty, J. C. (1957). Gonadotrophic cells of rat anterior pituitary gland and their relation to hormones produced. *Z. Zellforsch. Mikrosk. Anat.* **46**, 400.

Hotchkiss, R. D. (1948). A microchemical reaction resulting in the staining of polysaccharide structures in fixed tissue preparations. *Arch. Biochem.* **16**, 131.

Hunt, T. E., and Hunt, E. A. (1966). A radioautographic study of the proliferative activity of adrenocortical and hypophyseal cells of the rat at different periods of the estrous cycle. *Anat. Rec.* **156**, 361.

Hymer, W. C., and McShan, W. H. (1963). Isolation of rat pituitary granules and the study of their biochemical properties and hormonal activities. *J. Cell Biol.* **17**, 67.

Kobayashi, T., Kobayashi, T., Kigawa, T., Mizuno, M., Amenormori, Y. and Watanabe, T. (1967). Autoradiographic studies on H-leucine uptake by adenohypophyseal cells *in vitro*. *Endocrinol. Jap.* **14**, 69.

Koffler, D., and Fogel, M. (1964). Immunofluorescent localization of LH and FSH in the human adenohypophyis. *Proc. Soc. Exp. Biol. Med.* **115**, 1080.

Kracht, J. (1957). Zur Lokalisation der Hypophysenvorderlappenhormone. *Zentralbl. Allg. Pathol. Pathol. Anat.* **97**, 24.

Kragt, C. T., and Ganong, W. F. (1967). Pituitary FSH in female rats at various ages. *Fed. Proc., Fed. Amer. Soc. Exp. Biol.* **26**, 534 (abstr.) .

Kraicer, J., and Logothetopoulos, J. (1961). Distribution of gonadotropic and adrenocroticothrophic activity in the anterior pituitary gland of the male rat. *Endocrinology* **69**, 381.

Kurosumi, K., and Oota, Y. (1966). Corticotrophs in the anterior pituitary gland of gonadectomized and thyroidectomized rats as revealed by electron microscopy. *Endocrinology* **79**, 808.

Kurosumi, K., and Oota, Y. (1968). Electron microscopy of two types of gonadotrophs in the anterior pituitary gland of persistent estrus and persistent diestrus rats. *Z Zellforsch. Mikrosk. Anat.* **85**, 34.

McShan, W. H., and Hartley, M. W. (1965). Production, storage and release of anterior pituitary hormones. *Ergeb. Physiol., Biol. Chem. Exp. Pharmakol.* **56**, 264.

Midgley, A. R., Jr. (1963). Immunofluorescent localization human pituitary luteinizing hormone. *Exp. Cell Res.* **32**, 606.

Monroe, S. E., Rebar, R. W., Gay, V. L., and Midgley, A. R., Jr. (1969). Radioimmunoassay determination of luteinizing hormones during the estrous cycle of the rat. *Endocrinology* **85**, 720.

Nakane, P. K. (1970). Classification of anterior pituitary cell types with immunoenzyme histochemistry. *J. Histochem. Cytochem.* **18**, 9.

Nakayama, I., Nickerson, P. A., and Skelton, F. R. (1970). An electron microscopic study of the changes in ACTH and FSH producing cells during development of methylandrostendiol hypertension in the rat. *Amer. J. Pathol.* **58**, 377.

Novikoff, A. B., Essner, E., and Quintana, N. (1964). Golgi apparatus and lysosomes *Fed. Proc., Fed. Amer. Soc. Exp. Biol.* **23**, 1010.

Parlow, A. F. (1961). In "Human Pituitary Gonadotropins" (A Albert, ed.), p. 300. Thomas, Springfield, Illinois.

Parlow, A. F. (1968) . Species differences in regulation of pituitary gonadotropins. *Proc., Fed. Amer. Soc. Exp. Biol.* **27,** 269 (abstr.).

Pearse, A. G. E. (1948). Cytochemistry of the gonadotrophic hormones. *Nature (London)* **162,** 651.

Pearse, A. G. E. (1949). The cytochemical demonstration of gonadotropic hormone in the human anterior hypophysis. *J. Pathol. Bacteriol.* **61,** 195.

Pearse, A. G. E. (1950). Differential stain for the human and animal anterior hypophysis. *Stain Technol.* **25,** 95.

Pearse, A. G. E. (1953). Cytological and cytochemical investigations on the foetal and adult hypophsis in various physiological and pathological states. *J. Pathol. Bacteriol.* **65,** 355.

Perdue, J. F., and McShan, W. H. (1962). Isolation and biochemical study of secretory granules from rat pituitary glands. *J. Cell Biol.* **15,** 159.

Pomerantz, D. E., and Simmons, K. R. (1968). Immunofluorescent localization of bovine LH. *J. Histochem. Cytochem.* **16,** 205.

Purves, H. D., and Griesbach, W. E. (1951). The site of thyrotropin and gonadptropin production in the rat pituitary studied by McManus-Hotchkiss staining for glycoprotein. *Endocrinology* **49,** 244.

Purves, H. D., and Griesbach, W. E., (1954). The site of follicle-stimulating and luteinizing hormone production in the rat pituitary glands. *Endocrinology* **55,** 785.

Purves, H. D., and Griesbach, W. E., (1955). Changes in gonadotropes of rat pituitary after gonadectomy. *Endocrinology* **56,** 374.

Purves, H. D., and Griesbach, W. E. (1952). Functional differentiation in the acidophil cells and gonadotropic basophil cells of the rat pituitary. *Proc. Univ. Otago Med. Sch.* **30,** 27.

Ramirez, D. V., and McCann, S M. (1963). Comparison of the regulation of luteinizing hormone (LH) secretion in immature and adult rats. *Endocrinology* **72,** 452.

Rennels, E. G. (1957). Two tinctorial types of gonadotropic cells in the rat hypophysis. *Z. Zellforsch. Mikrosk. Anat.* **45,** 464.

Rennels, E. G. (1963). Gonadotropic cells of rat hypophysis. In "Cytologie d'Adenohypophyse" (J. Benoit and C. Da Lage, eds.) , pp. 128 and 201. Editions du Centre National de la Recherche scientifique. Paris.

Robyn, C. Y., Bossaert, P., Hukinout, O., Pasteels, J. L., and Herlant, M. (1964). *C. R. Acad. Sci.* **259,** 1226.

Rokhlina, M. L. (1940). *Dokl. Akad. Nauk SSSR* **27,** 504.

Romeis, B. (1940) Inkretorische organe. In "Handbuch der mikroskopischen Anatomie der Menschen" (F. von Möllendorff, ed.) , Vol. 6, Part 3, p. 79. Springer-Verlag, Berlin and New York.

Roos, J. (1968). Evolution des cellules gonadotropes prehypophysaires au cours du cycle oestral chez la ratte. *Z. Zellforsch. Mikrosk. Anat.* **84,** 372.

Roos, J. (1969). Action exercée par des agents antiovulatoires au stade prooestral du cycle ovarien, sur activité gonadotrope excrétrice des cellule β et Y de la pre hypophyse, chez la ratte. *Z. Zellforsch. Mikrosk. Anat.* **95,** 347.

Schleidt, J. (1914). Über die Hypophyse bei feminierten Mannchen und maskulierten Weibchen. *Zentralbl. Physiol.* **27,** 1170.

Severinghaus, A. E. (1937). Cellular changes in the anterior hypophysis with special reference to its secretory activties. *Physiol. Rev.* **17,** 556.

Shiino, M., Arimura, A., Schally, A. V. and Rennels, E. G. (1972). Ultrastructural observations of granule extrusion from rat anterior pituitary cells after injection of LH-releasing hormone. *Zellforsch. Mikrosk. Anat.* **128,** 152.

Smelser, G. K. (1944). Differential centrifugation of hormones in the central and peripheral zones of the bovine anterior pituitary gland. *Endocrinology* **34,** 39.

Smith, P. E. (1930). Hypophysectomy and a replacement therapy in the rat. *Amer. J. Anat.* **45,** 205.

Steelman, S. L., and Pohley, F. M. (1953). Assay of the follicle stimulating hormone based on the augmentation with human chorionic gonadotropin. *Endocrinology* **53,** 604.

Vanha-Perttula, T. (1966). Esterases of the rat adenohypophysis: Cellular localization and activity in relation to secretory functions. *Acta Physiol. Scand.* **69,** Suppl. 283 104 pp.

Wächtler, K., and Pearse, A. G. E. (1966). The histochemical demonstration of five lysosomal enzymes in the pars distalis of the amphibian pituitary. *Z. Zellforsch. Mikrosk. Anat.* **69,** 326.

Wolfe, J. M., and Brown, A. D. (1942). Action of diethylstilbestrol on cytological characteristics of anterior pituitaries of female rats, together with certain observations on the effects of castration. *Endocrinology* **31,** 467.

Yamada, K., and Sano, N. (1960). Electron microscopic observations of the anterior pituitary of the mouse. *Okajiimas Folia Anat. Jap.* **34,** 449.

Yoshimura, F., and Harumiya, K. (1965). Electron microscopy of the anterior lobe of the pituitary in normal and castrated rats. *Endocrinol. Jap.* **12,** 119.

Chapter 5

Adrenocorticotropes

Addison (1855) was the first to relate the symptoms of adrenal insufficiency to the adrenal glands. The relationship between the anterior pituitary and the adrenal cortex was not firmly established until the hypophysectomy and replacement therapy experiments by Smith (1930) and Grollman and Firor (1935).

LIGHT MICROSCOPY

Although the importance of the anterior pituitary gland in maintaining the adrenal cortex is known, the exact cell type responsible for the secretion of adrenocorticotropin (ACTH) has long been a point of disagreement. Basophils, acidophils, and chromophobes have each been implicated as the cellular source of ACTH.

Basophils

Kraus (1923, 1927) noted that in Addison's disease there was a scarcity of well-granulated basophils. A basophilic adenoma was associated with an unusual syndrome termed by Cushing (1932) as pituitary basophilism. Cushing's disease is now considered to be due to hypersecretion of adrenal corticoids, but Cushing believed it to be due to an abnormal adenomatous pituitary causing hyperplasia of the adrenal cortex. Crooke and Russell (1935) described hyalinized basophils in Cushing's disease.

Severinghaus (1938) described a case of Addison's disease and reported a depletion of basophils. Grollman and Firor (1935) reported vacuolated basophils in dogs suffering from adrenal insufficiency. Reese *et al.* (1939) observed the pituitary glands of rats following adrenalectomy and indicated decreased activity in certain acidophils and basophils but they noted that one population of basophils showed signs of high secretory activity. These basophils were believed to be the site of ACTH production. Koneff (1944) found that only basophils changed after prolonged ACTH treatment.

McLetchie (1942) and Mellgren (1945) described hyperfunctioning of basophils as a cause of Cushing's disease. Mellgren (1948) and Laqueur (1951) associated adrenogenital syndrome with a strong basophilic pituitary activity, and noted that subsequent adrenalectomy raised ACTH production. D'Angelo *et al.* (1948) reported an increase in the number of basophils in starve-stressed guinea pigs that exhibited adrenocortical hyperplasia. Sayers and Cheng (1949) found an 80% decrease in ACTH in pituitaries from adrenalectomized rats, which correlated with a decrease in basophils. These workers also observed an increase in ACTH associated with corticoid treatment. Taylor *et al.* (1949) and Tuchmann-Duplessis (1952) found that high inherent or injected levels of cortisone or desoxycorticosterone inhibited ACTH release and promoted storage of the hormone in the gland. Conversely in a hypocorticoid condition ACTH secretion was increased and the stores of hormone were depleted. Pearse (1953) and Purves (1956) reviewed the literature concerned with adrenal hypo- and hyperfunction and arthritis in man. They implicated a belief that changes in basophils are associated with these conditions.

Knigge (1955a,b, 1956, 1957) carried out extensive studies on cells presumed to produce ACTH. In his earlier publications he concluded that those cells which were stained by the protargol method produced ACTH because they changed during stress and they were related to cells that stain with PAS. In later work he noted that the number of basophils decreased and that those present displayed an altered cytology after adrenalectomy. He concluded that ACTH is produced by basophils. The work of Halmi and Bogdanove (1951), Kief and Schurmeyer (1956), and Dhom *et al.* (1962) strongly suggests that basophils, which stain with aldehyde fuchsin, secrete both ACTH and TSH.

Immunofluorescence data are also controversial. Marshall (1951) was the first to study the pituitary using immunofluorescent techniques. Marshall (1951), Pearse and van Noorden (1963), and Hess *et al.* (1968) have indicated that certain basophils are the source of ACTH.

Acidophils

The evidence pointing to acidophilic cells as the site of ACTH production is not extensive. Heinbecker and Rolf (1944) observed that when the nerve tracts and the median eminence were severed from the pituitary, the adrenals remained normal, the acidophils of the pituitary gland did not change, and the basophils were inactive. They concluded that ACTH is secreted by acidophils. When Finerty and Briseno-Castrejon (1949) performed unilateral adrenalectomies on rats, there was a compensatory hypertrophy in the remaining gland and a 12% increase in acidophils compared to the controls. They concluded that ACTH is secreted by acidophils. When Hess and Hall (1951) administered deoxycorticosterone to rats, the acidophils were markedly decreased. McCormack *et al.* (1951) reported that Cushing's disease is associated with an acidophilic adenoma of the pituitary gland.

Finerty *et al.* (1952) observed that acidophils of stressed rats showed an increased phospholipid content and, they believed, that this indicated active synthesis of ACTH. McNary (1957) failed to duplicate the findings of the Finerty group but noted that epinephrine increased phospholipid content of acidophils. He postulated that high titers of epinephrine stimulated ACTH release.

Stigliani *et al.* (1954) reported that acidophilic granules have ACTH activity, and Field (1958) observed a decrease in acidophils after adrenalectomy. They associated ACTH synthesis with acidophils. Herlant (1953) found that isolated granules 0.5–1.0 μ in diameter taken from sheep pituitaries contained ACTH activity. Immunofluorescence studies by Leznoff *et al.* (1962) and Kracht *et al.* (1965) suggest that acidophils are the site of ACTH synthesis.

Amphophils

Amphophilic cells stain with both acidic and basic dyes, and they have been suggested as the source of ACTH by Burt and Verlardo (1954), Burt *et al.* (1954), Russfield (1955), Ezrin *et al.* (1956), and Russfield *et al.* (1956). These observations were usually made under pathological conditions such as tumors. Amphophils have also been implicated as a source of STH, TSH, LTH, and the gonadotropins. These cells are usually polygonally shaped. They are partially degranulated basophils with an abundance of mitochondria showing acidophilic staining.

Chromophobes

Evidence that the chromophobe is the site of ACTH production can be traced back to the postulation by Lehman (1929) and Oka (1937) who noted an increase in chromophobe cells after adrenalectomy. Furth *et al.* (1953) studied ACTH secreting transplantable tumors in mice. The cells appeared to be chromophobes but upon closer observation they were shown to contain very small granules. Salassa *et al.* (1959) associated Cushing's disease with chromophobes. Although there was no direct experimental evidence Purves and Griesbach (1956) postulated that ACTH was secreted by chromophobes.

Farquhar (1957) suggested, on the basis of electron microscopic studies of the rat anterior pituitary, that angular chromophobelike cells containing microvilli which face a colloid-filled follicle may be the site of ACTH production. After cortisone treatment there was an increase in the follicular colloid, but after unilateral adrenalectomy the area of the follicle and the colloid decreased. Rennels (1964) substantiated, in part, the findings of Farquhar.

Corticotrope Granulation as Revealed by Tetrachrome Staining. Schelin *et al.* (1964) and Descourt and Herlant (1963) reported that the so-called chromophobe adenomas possessed granules that would stain by the Cleveland-Wolfe (1933) trichrome method, which was revised as the tetrachrome technique by Pasteels and Herlant (1963). Racadot and Herlant (1960). Racadot (1963a,b), and Quenum and Herlant (1964) identified corticotropes in cats based on studies of pathological and various physiological conditions. They found that the chromophobe adenoma consisted of cells containing many fine granules that stained by the tetrachrome method. Adrenalectomy and Metopirone administration produced an increase in granulation in some chromophobes. Hydrocortisone caused a decrease in these fine acidophilic granules. Racadot (1963a) administered amphenone to cats and noted that this adrenal cortical blocking agent also affects the thyroid gland and reproductive tract. Amphenone originally was reported to cause an increase in granulation of ACTH cells. Later it was discovered that this compound also caused milk secretion in cats and thereby indicated that it also stimulated granulation of mammotropes. Brozman (1967), using tetrachrome staining methods and immunofluorescence, substantiated the conclusion of the Herlant group that cells with fine granules produce ACTH. Dubois and Herlant (1968) using this tetrachrome method also identified ACTH cells in beef pituitaries.

Autoradiography has also been used to identify the ACTH cell. Siperstein (1963) and Knutson (1963, 1966a,b) conducted light autoradiographic studies in an attempt to identify the source of ACTH. In the study by Siperstein (1963) the corticotrope or adrenalectomy cell type was distinct from thyrotropes and gonadotropes. It was shown that after adrenalectomy small chromophobes differentiate into large chromophobes which have the highest protein turnover rate as determined by the uptake of tritiated glycine. On the basis of these results it was concluded that large chromophobes are the source of ACTH in the adrenalectomized rat. The work of Knutson (1966a,b) largely supports the results and interpretation of Siperstein. He noted that chromophobes increase in number and size up to 8 days after adrenalectomy but 14 days after the extirpation some basophils also were labeled. He interpreted this to mean that when the demand for ACTH is high not only do chromophobes secrete ACTH, but PAS positive basophils also synthesize the hormone. Even in the intact rat there is an increase in RNA synthesis in chromophobes after the animal is stressed.

In summary, the experimental evidence obtained from autoradiographic experiments indicates that the corticotrope is a large chromophobe. Based on tetrachrome staining this cell would be designated as a finely granulated acidophil.

ELECTRON MICROSCOPY

Using electron microscopy Siperstein and Allison (1965), Kurosumi and Kobayashi (1966), and Kurosumi and Oota (1966) have observed ACTH cells in sections from intact and adrenalectomized rat pituitaries. Siperstein and Allison (1965) made observations of pituitary glands from 7-day adrenalectomized rats. They reported the corticotrope was a large and irregularly shaped cell containing few granules and a vacuolated endoplasmic reticulum. The granules were about 200 mμ in diameter. Kurosumi and Kobayashi (1966) and Kurosumi and Oota (1966) observed corticotropes in sections from glands of intact and adrenalectomized rats. They found the greatest granulation in these cells 4 days after adrenalectomy. Seven days following adrenalectomy many ACTH cells were degranulated and degenerating. The ACTH granules exhibited variable electron densities and many had a dense core. These investigators estimated the granules to be 150–200 mμ in diameter. In further work they observed that corticotropes did not change after thyroidectomy or castration, indicating the corticotrope is a distinct cell type. Yamada and Yamashita (1967) studied the possible site of corticotropin synthesis in glands of mice and confirmed the identification of the corticotropes made by the earlier electron microscopists. When glutaraldehyde fixation was used, they found fewer ACTH granules with dense cores.

Corticotropes in Untreated Rats

Corticotropes are found extensively in the posteromedial region of the pituitary gland. There are more ACTH cells and greater granulation in

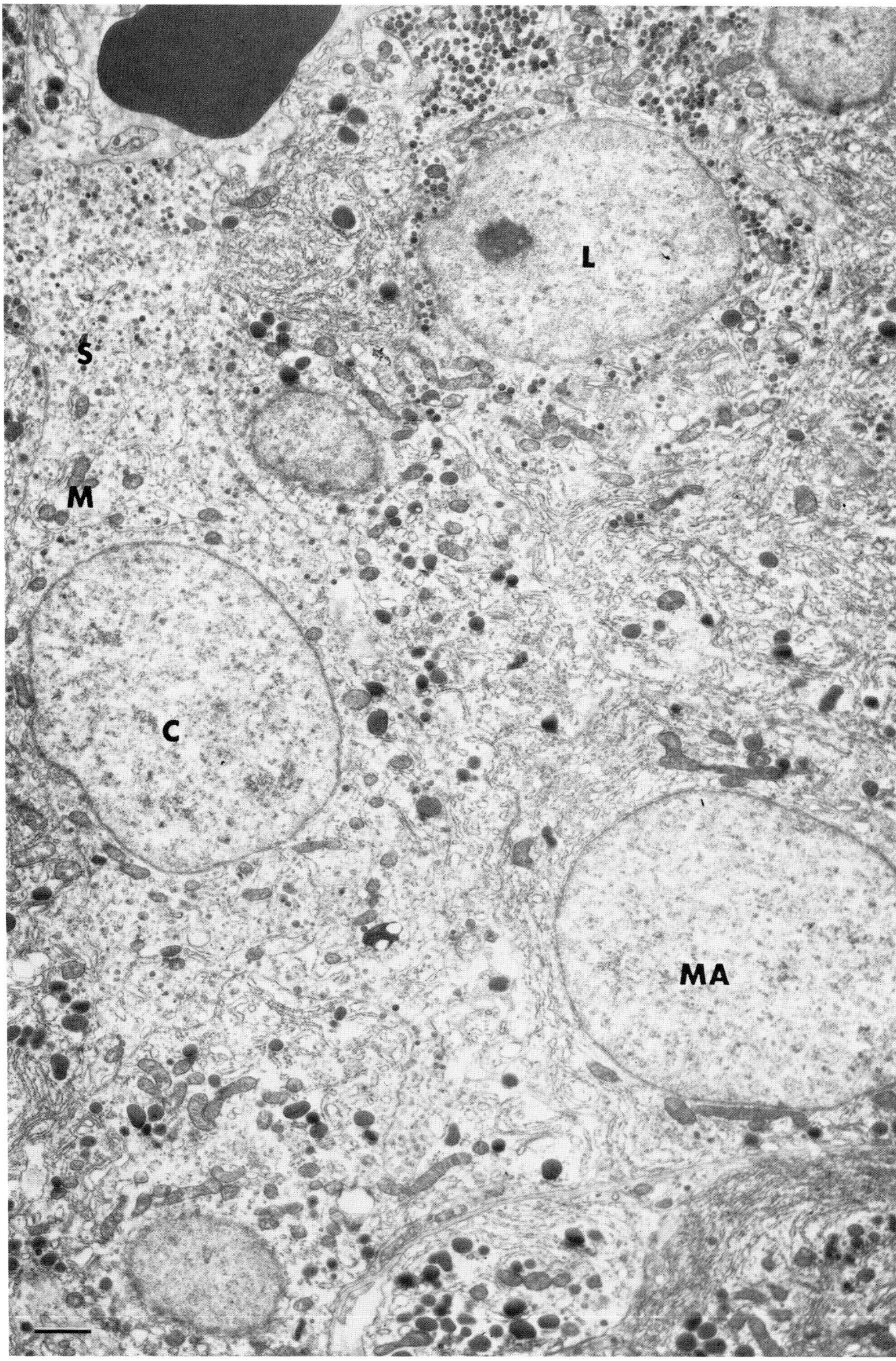

Fig. 5.1. A thin section from the anterior pituitary of a young male rat. A corticotrope (C), a mammotrope (MA), and a luteinizing hormone gonadotrope (L) are shown. The corticotrope is irregular in shape and well granulated (S). Short rod shaped mitochondria (M) and lysosomes are present. × 7000. (Veronal acetate buffered osmic acid)

the female than the male rat gland. Corticotropes are more abundant than thyrotropes and comprise about 2–4% of the pituitary cell population. The ACTH cells are large and irregular in shape (Fig. 5.1), but usually smaller than FSH gonadotropes. These corticotropes are often so irregular in form that they extend narrow cytoplasmic projections between other cells. They are usually found in clusters and are often associated with LTH and FSH cells.

The secretory granules of the corticotrope varied more in staining density than those in other cell types (Fig. 5.2). They ranged from either very electron dense to empty sacs showing little electron density. Some granules appear to be disrupted. In other cases they had a dense core surrounded by a clear space under the granule membrane. They appeared as though their contents either had shrunken away from the membrane or diffused from it. No other cell type had as many granules of this kind. The number of secretory granules in corticotropes is greater than in other cells containing small granules. The mean diameter of these granules was 106 mμ, and their range was 50–180 mμ.

The nucleus of the corticotrope is eccentrically located in the cell and usually rather irregular in shape. Golgi regions are numerous and encircle the nucleus, and consist of long, flattened saccules and vesicles. The endoplasmic reticulum is more often vesicular although on occasion cells could be seen with a lamellar endoplasmic reticulum. Ribosomes are sometimes attached to the endoplasmic reticulum but many ribosomes could also be found free in the cytoplasm, clustered in rosette-shaped figures. Centrioles with cilia projecting from the cell surface are often seen and lysosomes are numerous. Mitochondria are of different irregular sizes and shapes (Fig. 5.2) and are often aggregated around the Golgi regions. The mitochondrial cristae are sparse and irregular and the matrix is usually clear.

Corticotropes in Adrenalectomized Rats

Corticotropes showed greater morphological changes after adrenalectomy than any other cell type of the rat anterior pituitary gland. One day after adrenalectomy the corticotropes appear rather normal in cell outline. The Golgi apparatus and rough endoplasmic reticulum were much like they were in the intact rat (Fig. 5.3) but many of these corticotropes appeared degranulated.

After 2 days of adrenalectomy degranulation in corticotropes was more extensive (Fig. 5.4). Most of the organelles were like those in the intact rat; the only exception was an increase in lysosomes. Three days after adrenalectomy there was still a preponderance of poorly granulated

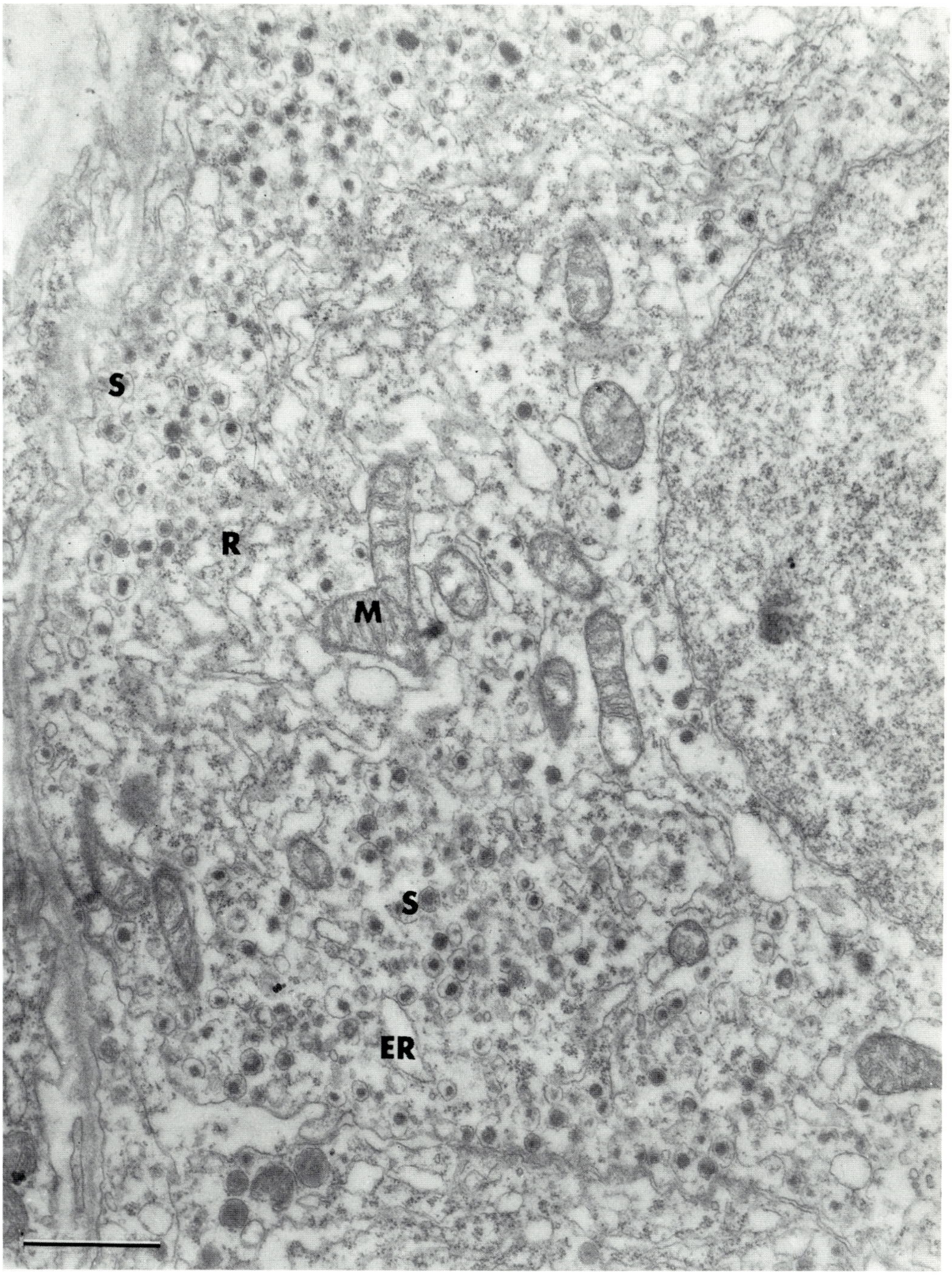

Fig. 5.2. A corticotrope from the pituitary of a male rat. This large, irregularly shaped cell contains secretory granules (S) of various electron densities. Membranes loosely surround the granules. Endoplasmic reticulum (ER) is rather extensive and contains ribosomes attached to the saccules; some ribosomes (R) are free in the cytoplasm. Mitochondria (M) of varying shapes with scattered cristae are present. × 20,000. (Veronal acetate buffered osmic acid

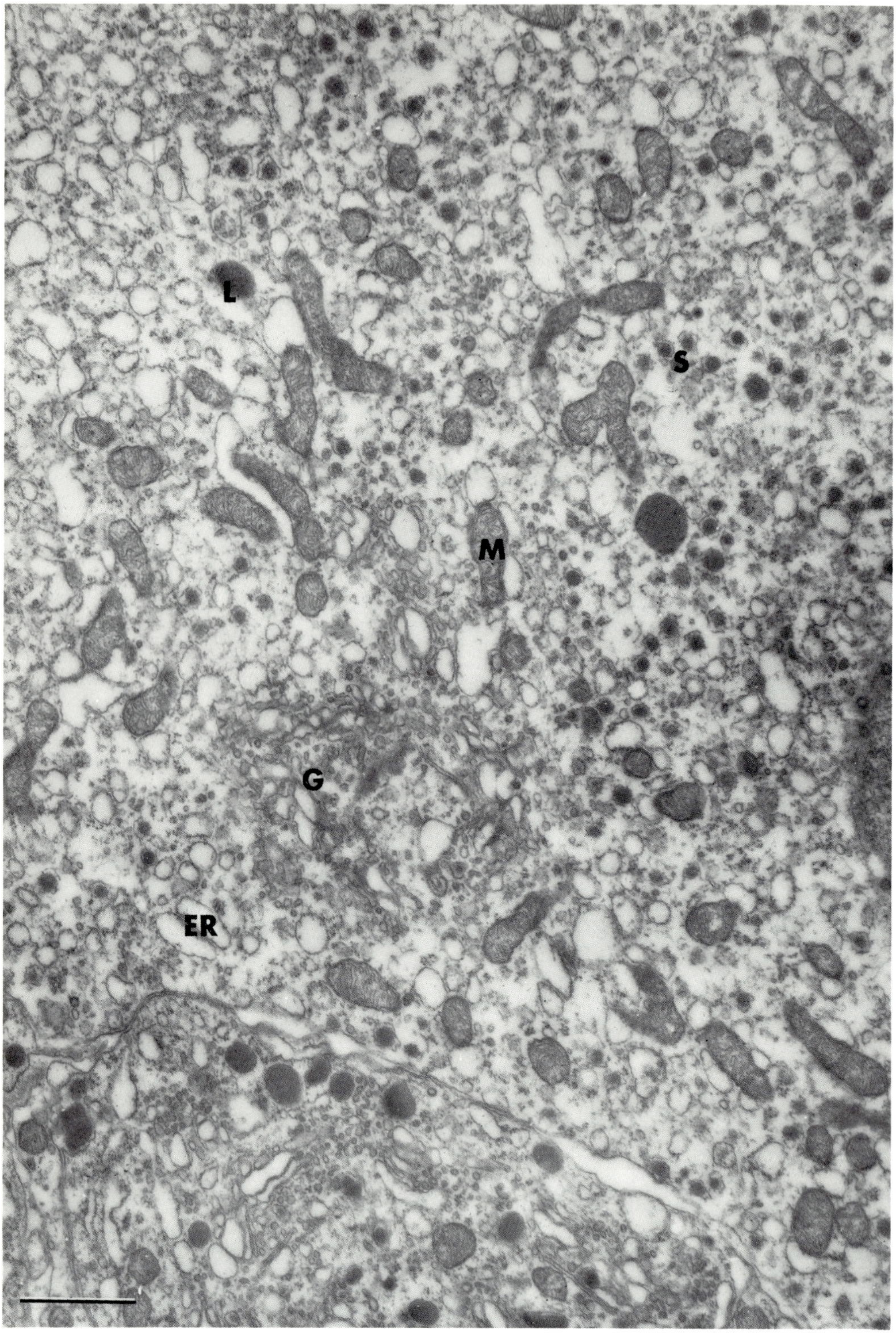

Fig. 5.3. A corticotrope of the pituitary 1 day after adrenalectomy. The endoplasmic reticulum (ER) appears more like hypertrophied saccules usually devoid of attached ribosomes. A Golgi complex (G), mitochondria with dense cristae (M), lysosomes (L), and secretory granules (S) are indicated. × 18,000. (Veronal acetate buffered osmic acid)

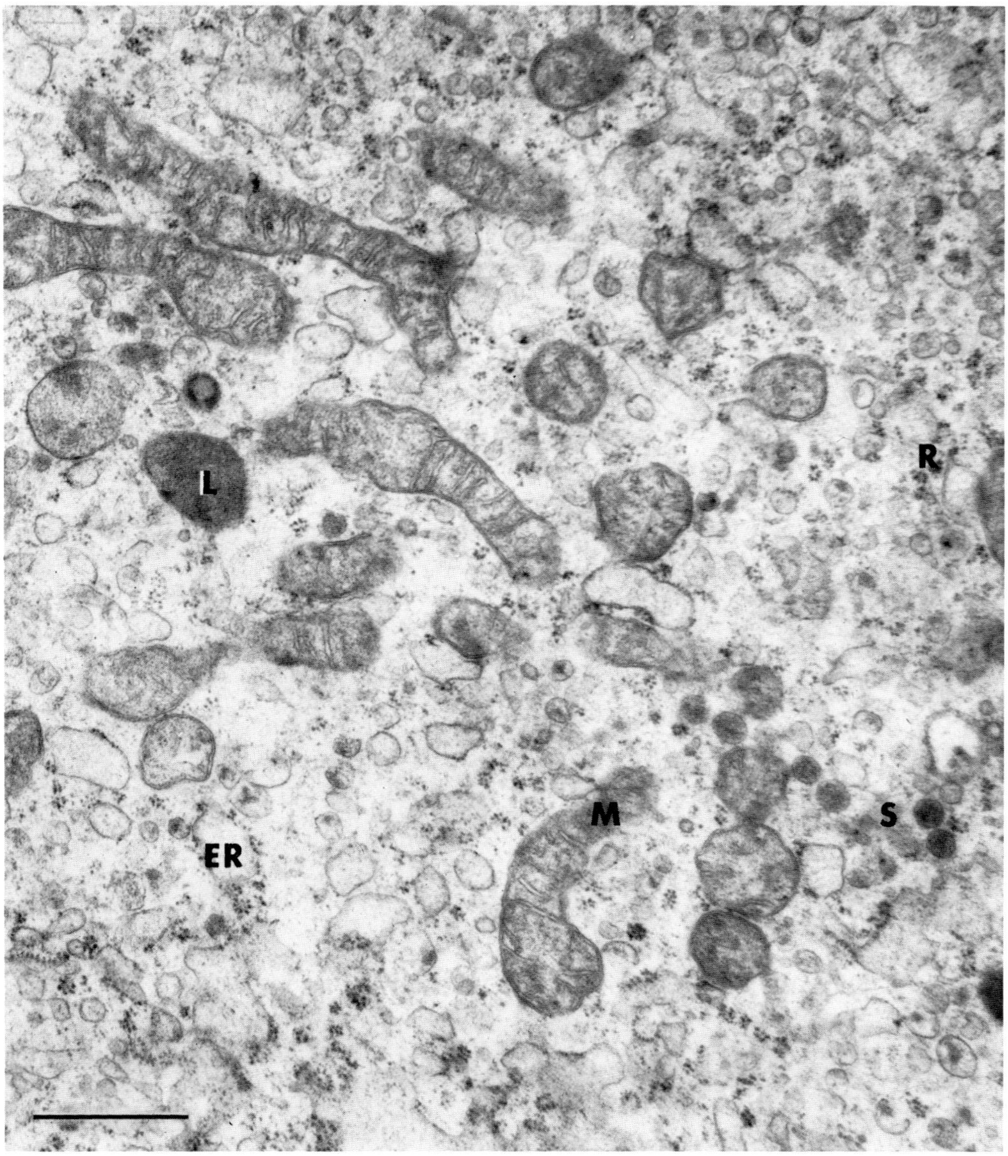

Fig. 5.4. A corticotrope 2 days after adrenalectomy. Vesicular endoplasmic reticulum (ER) is hypertrophied but has few attached ribosomes. Well-defined clusters of ribosomes (R) are present. The granules (S) are in different stages of degranulation. Lysosomes (L) and mitochondria (M) are shown. × 21,000. (Veronal acetate buffered osmic acid)

cells, but the endoplasmic reticulum of many corticotropes was hypertrophied and there was an increase in ribosomes. The Golgi complexes were hypertrophied. Four days after adrenalectomy most cells appeared densely granulated. Many granules in various stages of formation were observed in the enlarged Golgi complex (Fig. 5.5). At this stage there was also some growth and differentiation of the chromophobe cells. The chromophobes surrounding the pseudofollicles were also enlarged.

Five days following adrenalectomy there was an increase in corticotropes which were enlarged and irregular in shape (Fig. 5.6). The granulation of the corticotropes was greatest at this time. The rough endoplasmic reticulum had greatly enlarged from a vesicular to a lamellar type characterized by parallel arrays of flat cisternae (Fig. 5.7). The Golgi complexes were active; they had hypertrophied saccules, which were apparently forming more secretory granules. There also appeared to be an increase in lysosomes in these cells. Seven days after adrenalectomy some of the corticotropes appeared degranulated (Fig. 5.8). The Golgi complexes and endoplasmic reticulum were no longer as greatly dilated as before, and there were more free ribosomes in the cytoplasm. Since there were fewer granules in each cell of this type and those present were not electron dense, it was difficult to detect these cells at this time. Many of these adrenalectomy cells now appeared quite similar to large chromophobe follicle cells. There was an increase in number of these cells and of cilia and microvilli extending into the follicular area. The pituitary cleft that separates the pars intermedia from the pars distalis was greatly hypertrophied (see Chap. 10).

Fourteen days after adrenalectomy there were fewer corticotropes than at day 5. These large, irregular cells contained many granules, swollen mitochondria with broken cristae, and a hypertrophied endoplasmic reticulum which was intermittently studded with ribosomes. Thirty to 60 days following adrenalectomy (Fig. 5.9) there were still a few corticotropes but they were often degranulated. The endoplasmic reticulum was still hypertrophied but there were more free ribosomes. Most mitochondria were greatly swollen and had broken cristae. Lysosomes and fatty droplets had increased in some corticotropes. Chromophobes tended to predominate, and the hypophyseal cleft was still hypertrophied at this time.

The action of adrenal cortical blockers, adrenal steroids, ACTH and α-ethyltryptamine on corticotrope ultrastructure is given in the following section.

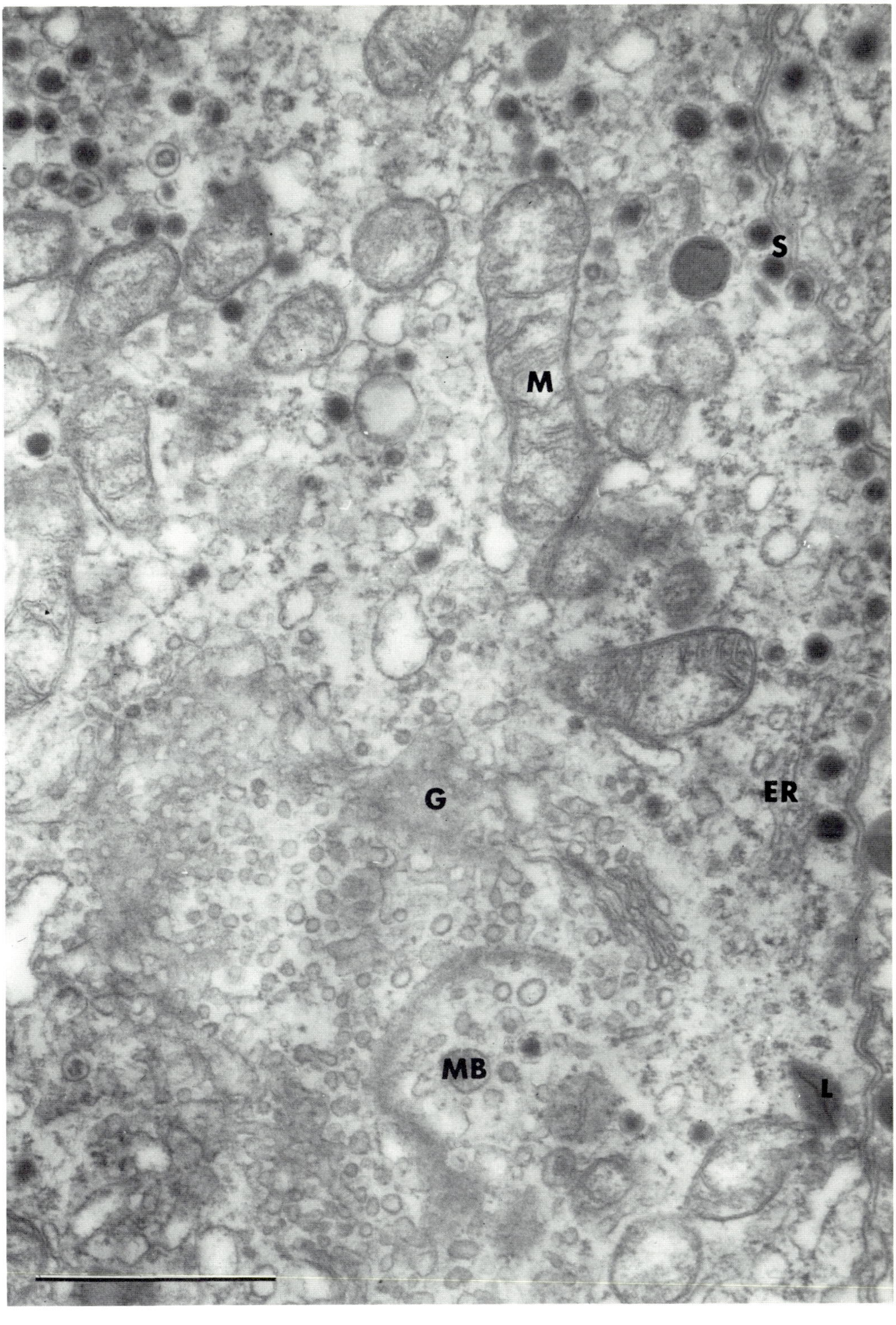

Fig. 55. A cortocotrope of a male rat pituitary 4 days after adrenalectomy. An extensive Golgi complex (G) and evenly electron dense secretory granules (S) characterize this cell. Endoplasmic reticulum (ER) is often studded with ribosomes. Mitochondria (M) containing interrupted scattered cristae, a lysosome (L) and a microbody (MB) are also present. × 37,500. (Veronal acetate buffered osmic acid)

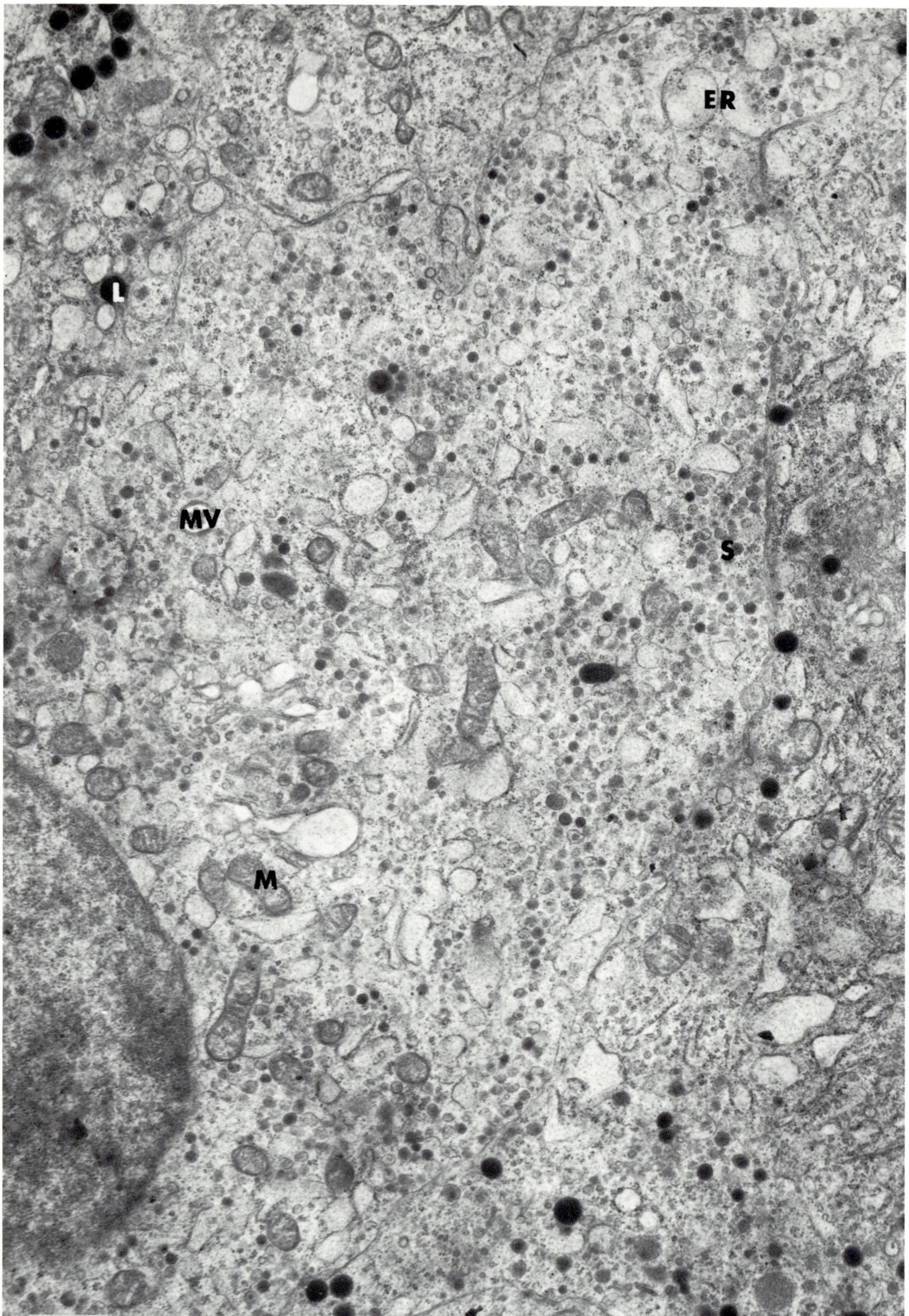

Fig. 5.6. A typical corticotrope from a pituitary of a rat adrenalectomized for 5 days. A hypertrophied, well-granulated corticotrope is shown. It is characterized by a great abundance of granules (S) of varying electron density, several lysosomes (L) and multivesicular bodies (MV). Mitochondria (M) surround the nucleus and vesicles of endoplasmic reticulum (ER) with attached ribosomes are present. × 12,000. (Veronal acetate buffered osmic acid)

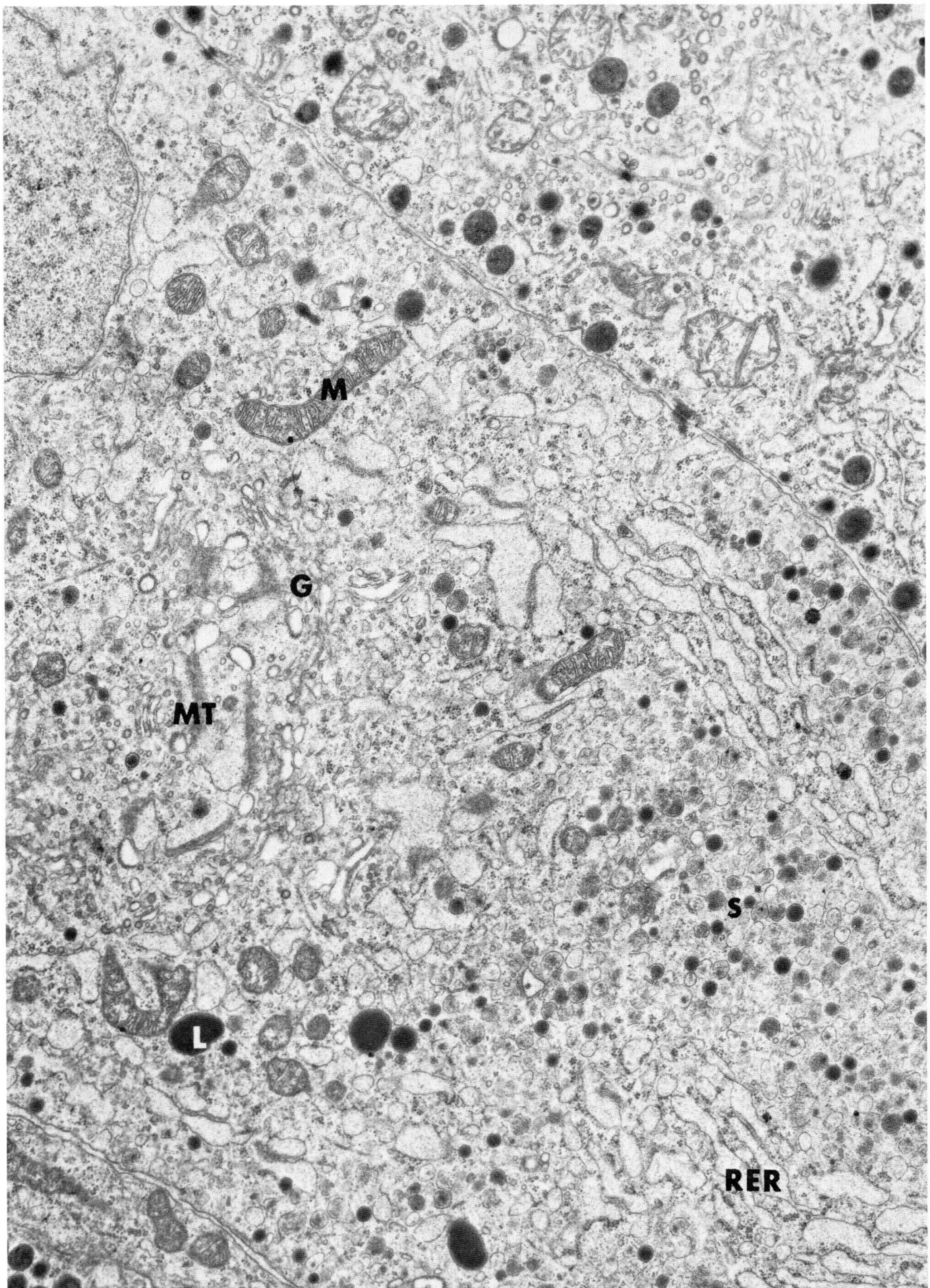

Fig. 5.7. A corticotrope of a male rat pituitary gland after 5 days of adrenalectomy. This part of the corticotrope exhibits an abundance of granules (S) of varying electron density and rough endoplasmic reticulum (RER) that is more lamellar in appearance. Lysosomes (L), round and rodlike mitochondria (M), a Golgi complex (G) and microtubules (MT) can also be seen in this cell. × 12,500. (Veronal acetate buffered osmic acid)

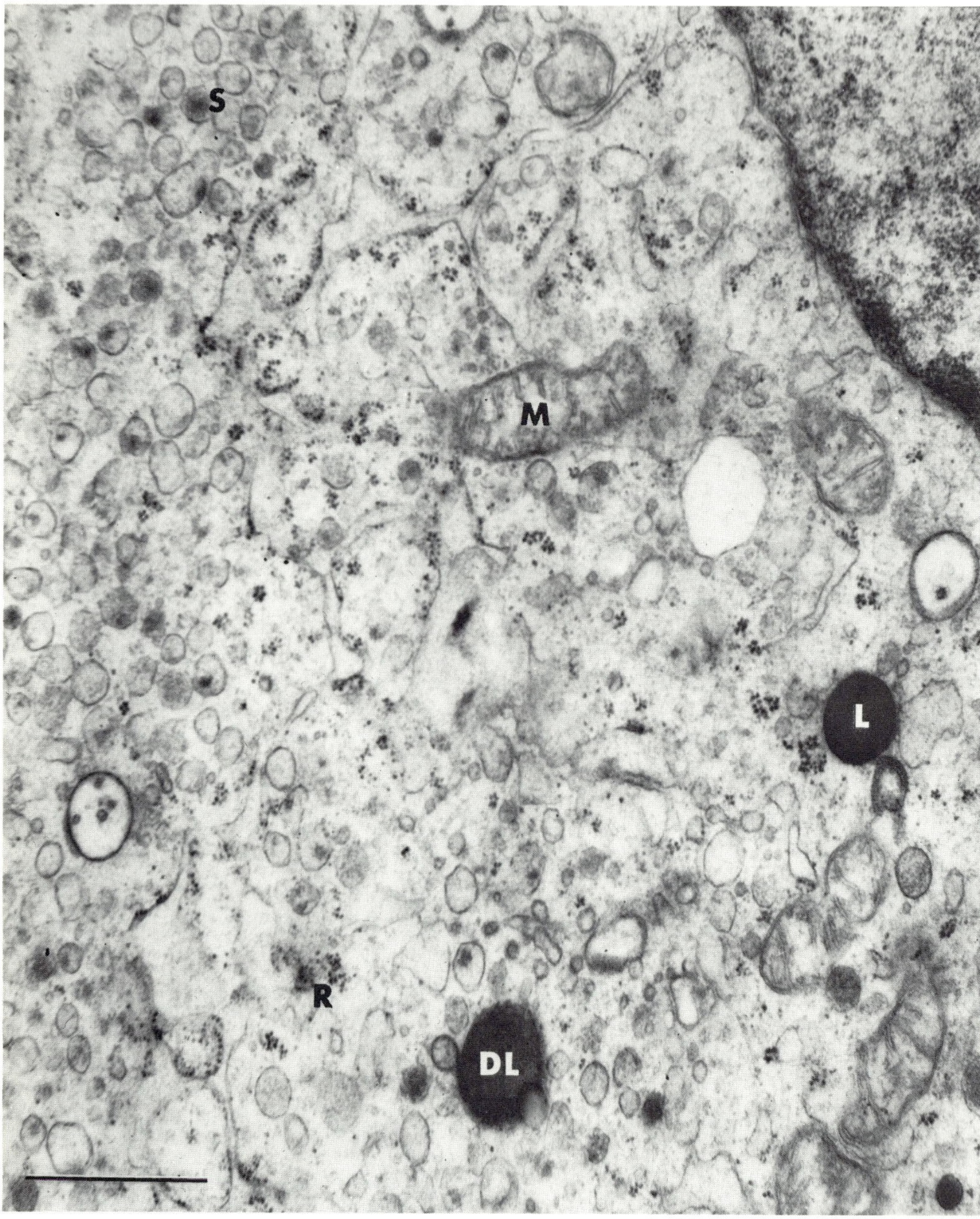

Fig. 5.8. A section through part of a corticotrope in the pituitary of a male rat adrenalectomized for 7 days. This corticotrope shows granules (S) in various stages of degranulation some of which appear as empty vesicles. A lysosome (L) and a lysosome absorbing a granule (DL) are present with mitochondria that appear to have disoriented cristae (M). There are scattered rosette shaped ribosome complexes (R). × 25,000. (Veronal acetate buffered osmic acid)

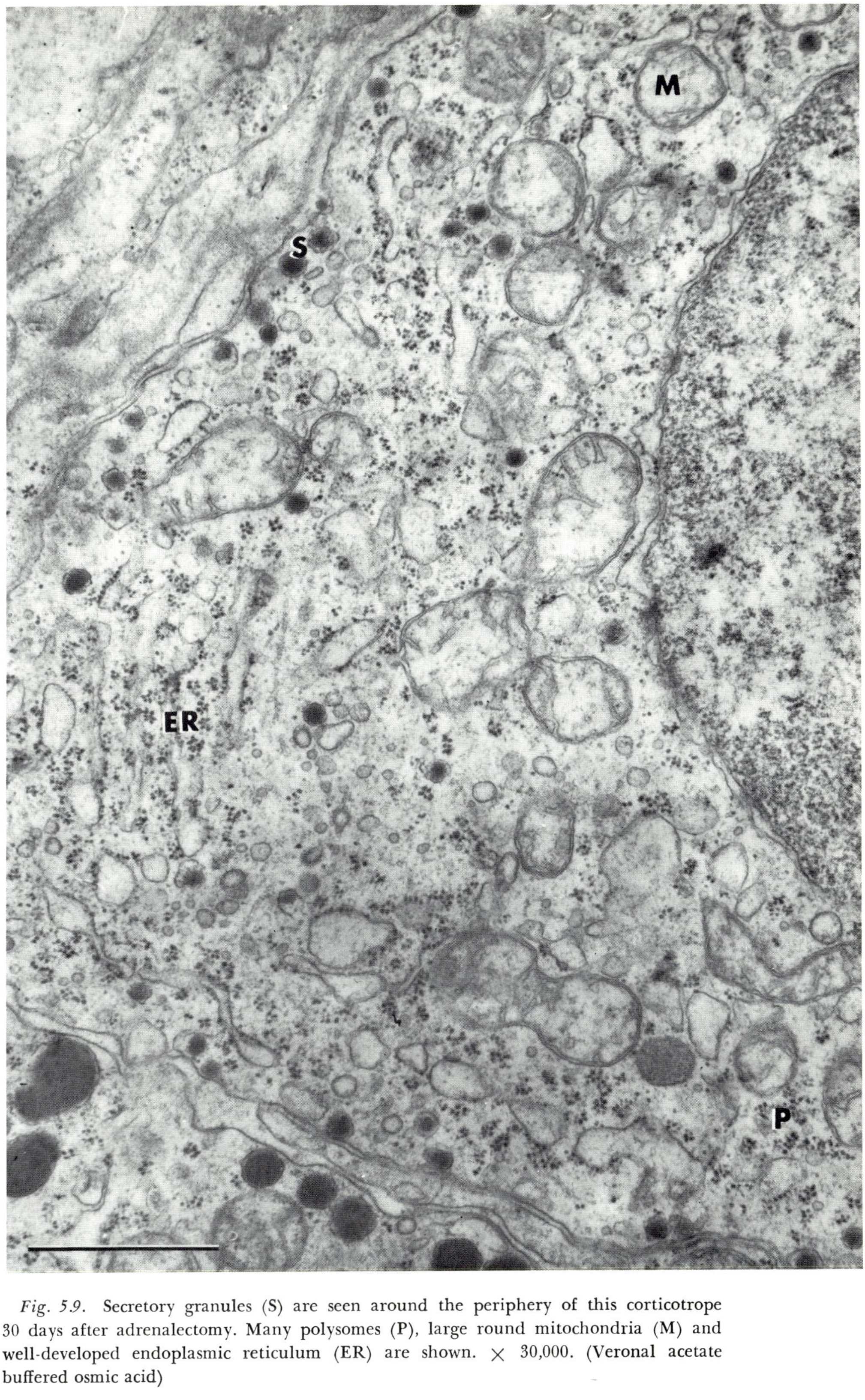

Fig. 5.9. Secretory granules (S) are seen around the periphery of this corticotrope 30 days after adrenalectomy. Many polysomes (P), large round mitochondria (M) and well-developed endoplasmic reticulum (ER) are shown. × 30,000. (Veronal acetate buffered osmic acid)

METOPIRONE

The compound 2-methyl 1-2-bis-(3-pyridyl)-1-propanone (Metopirone, methopyropone, SU-4885) was synthesized by Allen (1950) and first introduced by Chart *et al.* (1958). It has been shown by Liddle *et al.* (1958) in man and Jenkins *et al.* (1958, 1959) in the dog that Metopirone is a specific 11β-hydroxylase inhibitor at a low dose and at a high dose is a general inhibitor of adrenal corticoid synthesis.

Gold *et al.* (1959) measured the ability of the anterior pituitary to release ACTH by determining the level of 17α-OH compounds such as 17α,21-dihydroxy-4-pregnene, and 3,20-dione-11-deoxycortisol in the plasma and urine after blocking cortisol release with a large dose of Metopirone. Waxman *et al.* (1961) showed that Metopirone administration led to complete arrest of cortisol synthesis in man.

Dominguez and Samuels (1963) and Williamson and O'Donnell (1967) performed *in vitro* experiments to determine the type of inhibition of 11β-hydroxylation that was produced by Metopirone. They found that it acts as a competitive inhibitor. It was also found that a competition does not exist between 21-hydroxylase and 17-hydroxylase for the common substrate progesterone indicating that metopirone affects cortisol and not corticosterone synthesis.

In addition to inhibition of 11β-hydroxylation, Metopirone has been shown by Kraulis and Birmingham (1963) to inhibit 18-hydroxylation and Griffiths (1963) reported an inhibition of 19-hydroxylation. The inhibition of 18-hydroxylation means the conversion of progesterone to aldosterone by way of 18-hydroxycorticosterone is reduced by 50%. Finally, Cheng and Carballeira (1969) have shown that Metopirone exerts an inhibition of hydroxylation at the C-20 position of the steroid nucleus.

Janssens (1965) has observed that not only does Metopirone interfere with steroid synthesis but also with enzymes involved in urea formation in *Xenopus laevis*. Steenburg (1965) believed that Metopirone acted directly on the pituitary by influencing ACTH release. Herlant (1964) reported that Metopirone stimulated corticotropes in cat pituitaries. Farmer (1961) postulated that this compound acts at the pituitary level in man. Katsuki *et al.* (1967) showed that inhibition of the adrenal cortex by Metopirone was impaired after the anterior hypothalamus of dogs was lesioned. A functional dichotomy is proposed by these workers for the hypothalamus and its regulatory mechanism of ACTH secretion. Garweg *et al.* (1969) have shown that Metopirone affects other parts of the central nervous system besides the hypothalamus.

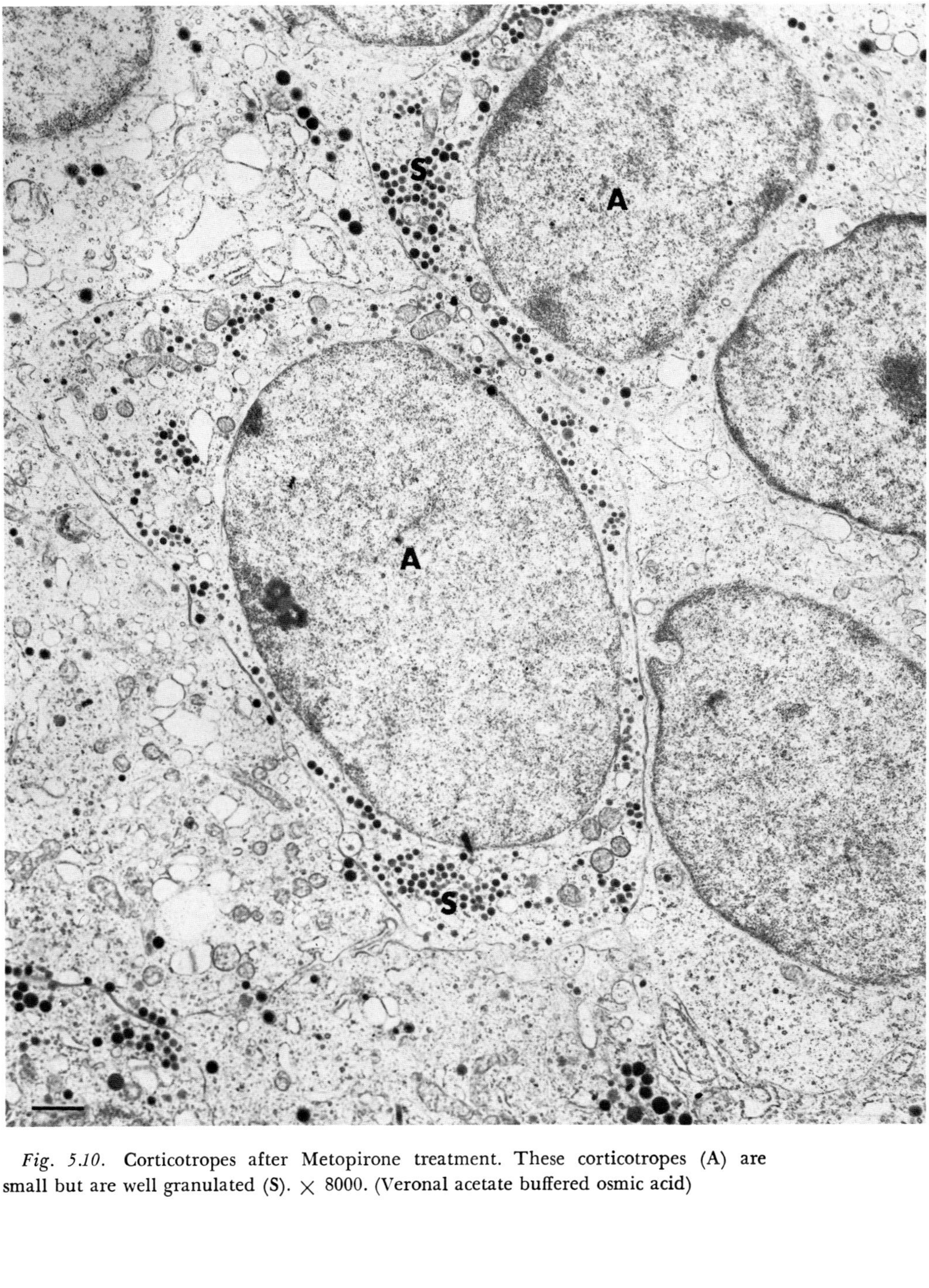

Fig. 5.10. Corticotropes after Metopirone treatment. These corticotropes (A) are small but are well granulated (S). $\times$ 8000. (Veronal acetate buffered osmic acid)

It is clear that Metopirone inhibits corticoid synthesis which prevents negative feedback to the pituitary. There is some fragmentary evidence that Metopirone may act either through the hypothalamus or directly on the pituitary.

Changes in Corticotropes after Metopirone Administration

The administration of Metopirone produced the following cytological changes. Some corticotropes appeared smaller but evenly granulated and had a well-developed Golgi complex (Fig. 5.10). Other corticotropes exhibited a hypertrophied endoplasmic reticulum with fewer attached ribosomes, a well-developed Golgi complex and fewer granules (Fig. 5.11). The changes in these may be interpreted to mean that the hormone was being produced but not being stored in the granules. The mitochondria in corticotropes were usually normal in size but the cristae were irregular or broken.

Because of the change in ultrastructure of the pituitary gland after Metopirone treatment there is evidence that this compound may act on the pituitary as well as the adrenal cortex.

AMPHENONE

The earliest amphenone-type compound was a substituted desoxybenzoin, 2,2-bis (*p*-chloroplenyl) -1,1-dichloroethane (Amphenone B) which was synthesized by Allen and Corwin (1951). Amphenone B and various derivatives have been implicated as inhibitors of adrenal corticoid synthesis. Two more widely used analogs are 3,3-bis-(*p*-aminophenyl)-2-butanone and 1,1-bis-(*p*-aminophenyl)-1-methyl propanone. These compounds were synthesized by Bencze and Allen (1957).

Hertz and Tullner (1950) showed that amphenone produces progestational effects on the end organs since these effects occur even in hypophysectomized rats. Hogness *et al.* (1953) reported that amphenone is an antithyroid compound and also an adrenal stimulant. Hertz *et al.* (1955) and Vilar and Tullner (1959) reported that although amphenone causes stimulation of the adrenal and also the pituitary in rats, it blocks adrenal steroid production in dogs. Nichols and Hennigar (1957) reported this compound might act on the pituitary to inhibit ACTH release. Rosenfield and Bascom (1956) showed that amphenone in *in vitro* culture inhibited 11β-, and 17α-, and 21-hydroxy enzymatic steps in steroid biosynthesis in the calf adrenal gland. This inhibition was reversible if a great amount of ACTH was added to the system. Kibelstis and Ferguson

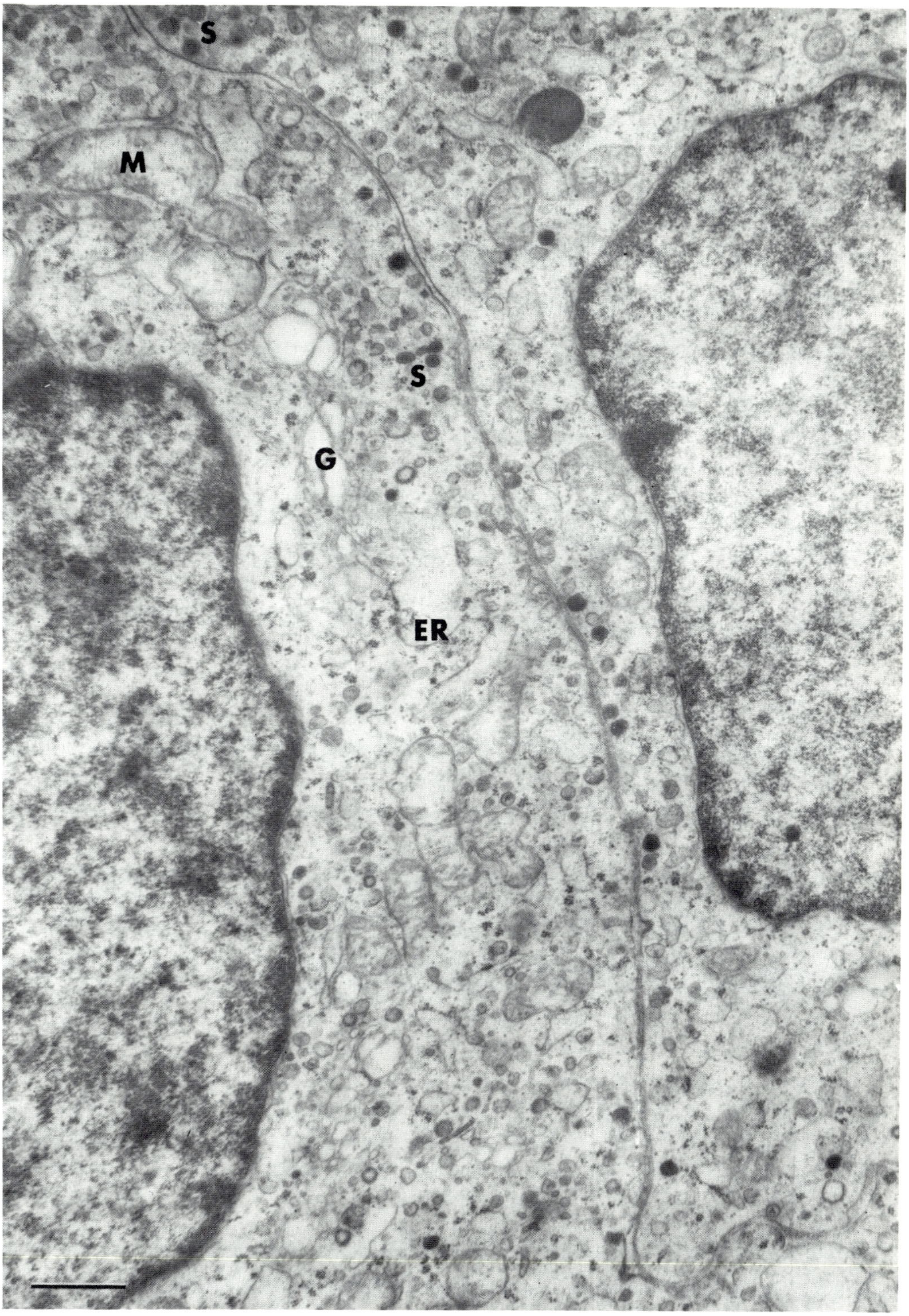

Fig. 5.11. Two corticotropes after Metopirone treatment. These cells have few granules (S), a hypertrophied Golgi complex (G), swollen mitochondria (M), and several cisternae of endoplasmic reticulum (ER). × 15,000. (Veronal acetate buffered osmic acid)

(1964) established that amphenone exerts its inhibition by blocking the conversion of 20α-hydroxycholesterol to pregnenolone.

Racadot and Herlant (1960) administered large doses of amphenone into cats and found an increase in adrenal weight and an increase in granulation of corticotropes. However, Racadot (1963a,b) reported that the 1960 findings were in error because the cats were lactating and many of the well-granulated cells were mammotropes and not corticotropes. Hohn and Sarker (1964) reported development of mammary alveolae due to amphenone administration even in adrenalectomized and castrate female rats. In intact animals they found an increase in adrenal and thyroid weight after amphenone administration.

It appears that amphenone influences the thyroid, mammary glands, and adrenals, and since certain tropic hormones govern these glands, it may also inhibit the anterior pituitary gland. There are species differences and many discrepancies associated with amphenone administration.

Ultrastructural Changes of Corticotropes after Amphenone Administration

The cellular picture of the pituitary after amphenone administration was quite similar to that of the pituitaries of rats treated with Metopirone. The central part of the gland appeared more like the controls than did the peripheral regions that are poorly vascularized. The corticotropes were rather poorly granulated (Fig. 5.12). The nuclei were more irregular, and some of the mitochondria were swollen such that the cristae were almost lacking. The endoplasmic reticulum was hypertrophied and the number of ribosomes were increased.

Amphenone appeared to act on the pituitary. It affected not only the fine structure of the corticotropes but also the cellular integrity of the entire gland. This treatment resulted in partial degranulation of some of the corticotropes similar to that which occurred 2 days after adrenalectomy.

Discussion of Metopirone and Amphenone Studies

Light microscopy studies of pituitaries of cats treated with amphenone and Metopirone were performed by Racadot and Herlant (1960) and Racadot (1963a,b), but they used unphysiological doses of these compounds. They observed hypertrophy of corticotropes and increased granulation after treatment with these blocking agents. They also noted stimulation of mammotropes after amphenone treatment. Quenum (1964) reported hypertrophy of corticotropes in Gambian rats treated with Metopirone.

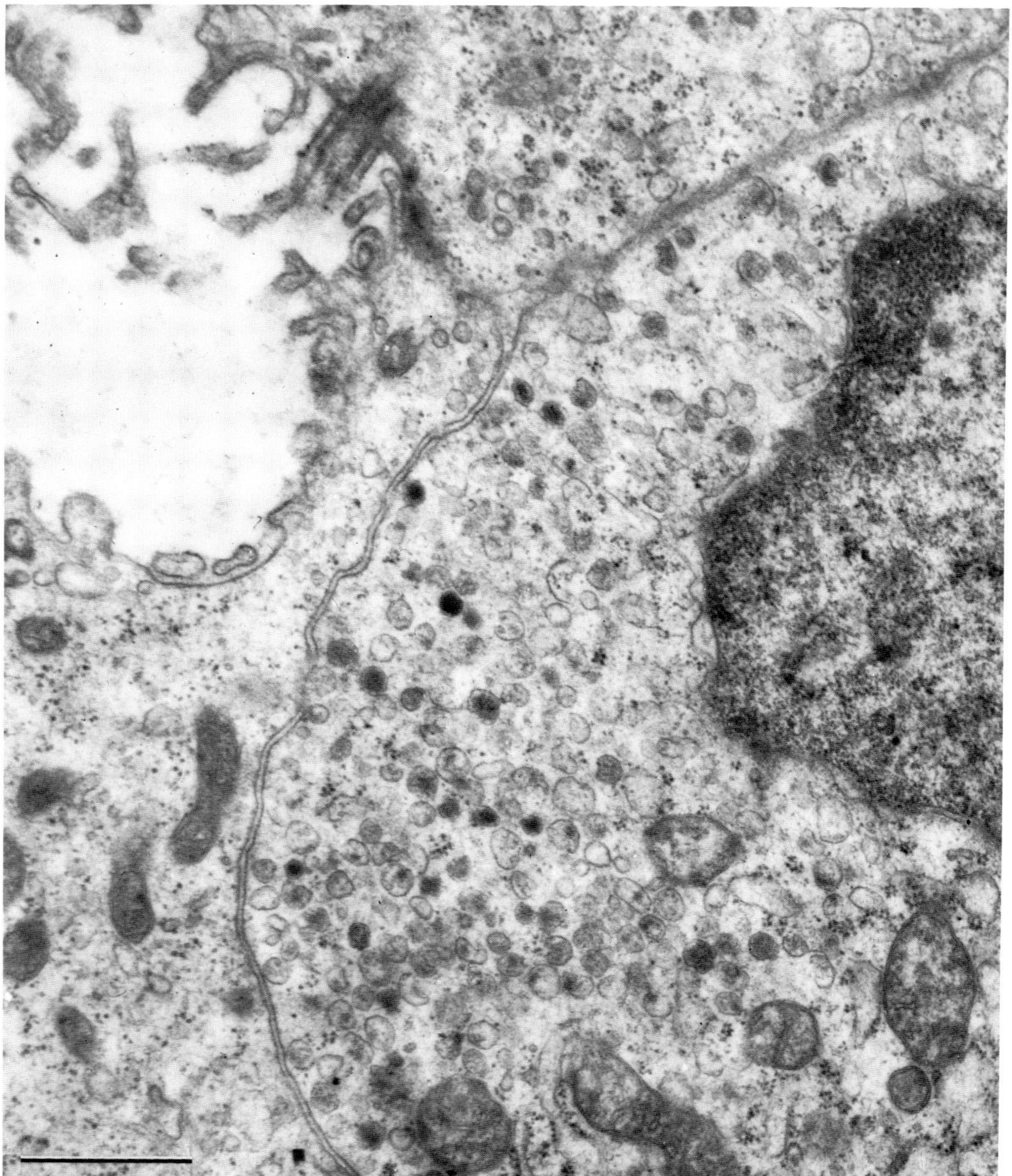

Fig. 5.12. A corticotrope after amphenone treatment. Granules in various stages of degranulation are shown. A follicle and follicle cell is shown at the left. × 23,450. (Veronal acetate buffered osmic acid)

In our studies these agents did not cause an increase in size or granulation of ACTH cells. The corticotropes in these studies exhibited different degrees of degranulation. However, cisternae of the endoplasmic reticulum were sometimes hypertrophied after the Metopirone treatment.

In our studies, especially after amphenone treatment, the pituitary appeared somewhat abnormal. Nuclear inclusions, greatly invaginated nuclei, and swollen mitochondria were observed. The fine structure of these pituitaries from treated rats was similar to that of pituitary tumors reported by Bernhard (1958) and Schelin *et al.* (1964) in the rat, and Krishan *et al.* (1967) in the hamster. It appears that these compounds act directly on the pituitary as well as on the adrenal cortex. This action has been postulated by Steenburg (1965).

DEXAMETHASONE

Dexamethasone is a synthetic steroid that is 28 times more potent than hydrocortisone. It was first used by Ringler and Brownfield (1960) as a pituitary blocking agent. Slusher (1965) reported that it is a more effective blocking agent of ACTH release than are other corticoids. When dexamethasone was infused into dogs for 4 hours, it suppressed the adrenal cortex. After 12 hours it reduced the pituitary store of ACTH as reported by Gold *et al.* (1961). Kendall (1961) observed that when it was administered subcutaneously maximum ACTH suppression occurred in 4 hours.

De Wied (1962) conducted extensive studies to determine the site of action of dexamethasone. Male rats were given this compound 16 hours before exposure to noxious stimuli. The adrenal response to ACTH was not affected by dexamethasone indicating it did not act on the adrenal. Dexamethasone did not inhibit ACTH secretion if the median eminence was lesioned, but otherwise secretion was inhibited. It was concluded that dexamethasone exerts its action by blocking ACTH release from the anterior pituitary gland.

Purves and Sirett (1965) administered dexamethasone to the drinking water of rats and reported complete corticotropin suppression. On the next day the ACTH preparation to be analyzed was administered by intrajuglar injection of doses as minute as 0.0005–0.002 IU and 18 minutes later a blood sample was taken and read fluorimetrically. This provided the basis for the development of a quick sensitive method for ACTH assay.

Recently, Arimura *et al.* (1967) used a combination of dexamethasone, Nembutal, and morphine as an assay system for corticotropin-releasing factor. Davidson and Feldman (1967) studied the possible inhibition of corticotropin secretion by selective implantation of dexamethasone. The conclusions were that either forebrain structures remote to the hypothalamus are sensitive to corticoid negative feedback or that perhaps the dexamethasone diffused from the forebrain to the hypothalamus causing ACTH inhibition.

Russell *et al.* (1969) reported that the release of corticotropin from the pituitary can be inhibited *in vitro* by dexamethasone which suggests that the anterior pituitary may be the site for corticosteroid negative feedback. This view has been substantiated by Arimura *et al.* (1969) since dexamethasone suppressed the action of CRF at the pituitary level. This blockage of CRF requires a process that involves DNA-dependent RNA synthesis.

It can be concluded that dexamethasone administration feeds back on the hypothalamus and/or pituitary and thus inhibits corticotropin secretion by the pituitary gland.

Ultrastructure of Corticotropes After Dexamethasone Treatment

This powerful synthetic adrenal cortical steroid appeared to cause an inhibition of corticotropes. Very few corticotropes were found in pituitaries from rats treated with dexamethasone. The few corticotropes present were smaller and appeared to be degenerating. The Golgi complex was atrophic or not observed, and the endoplasmic reticulum was fragmentary with few ribosomes. The mitochondria were swollen and usually lacked cristae.

In our study, administration of dexamethasone to rats caused corticotropes and other pituitary cells to undergo the following changes: swollen mitochondria often devoid of cristae, a vacuolated cytoplasm, degranulation, and an increase in lysosomes (Fig. 5.13).

The increase in lysosomes in pituitary cells probably indicates that pathological changes were taking place as has been shown by Novikoff and Essner (1962) and de Duve and Wattiaux (1966) in other organs. Swift and Hruban (1964) have also shown micrographs in which the organs of rats treated with various noxious compounds were greatly infiltrated with lysosomes. They concluded that lysosomes were induced by these noxious stimuli and that lysosomes may cause certain parts of the cytoplasm to be resorbed. It is clear that dexamethasone caused atrophy of pituitary corticotropes.

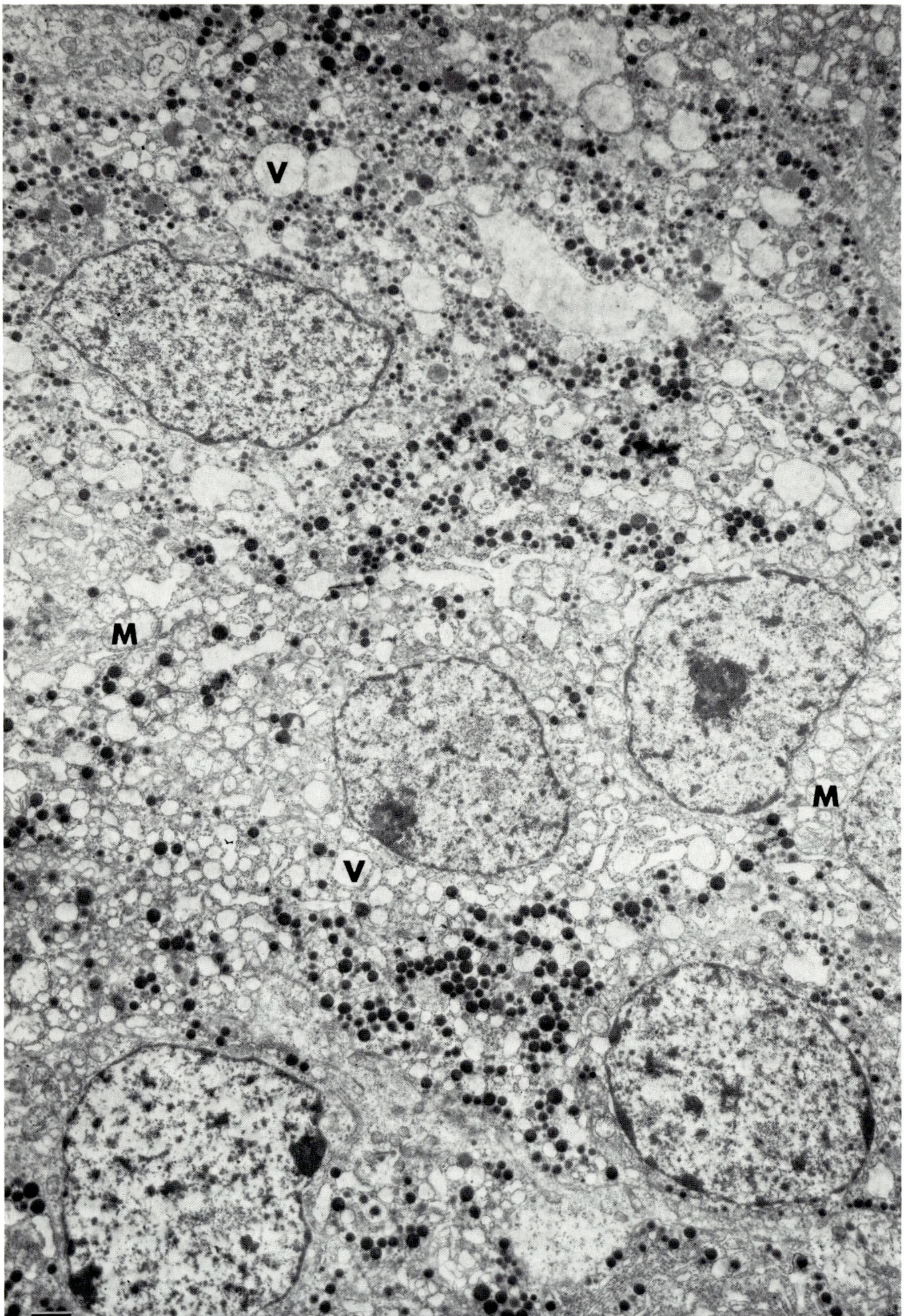

Fig. 5.13. A section of a pituitary from the rat after dexamethasone treatment. Great vacuolated areas (V), swollen mitochondria (M), a degranulation and lack of cell membranes are characteristic of the cytology of the pituitary gland after this treatment. Cell types are difficult to identify in this micrograph. $\times$ 5000. (Veronal acetate buffered osmic acid)

ADRENAL CORTICAL STEROIDS

It is dogma in endocrinology that pituitary hormones act upon and stimulate target organs to secrete hormones and that these target organs, like the adrenal cortex, secrete adrenal corticoids which act on the pituitary to regulate or inhibit ACTH secretion. (Turner and Bagnara, 1971).

Ingle and Kendall (1937) noted that cortisone administration induced adrenocortical atrophy in experimental animals. Sayers and Sayers (1947) showed that the secretion of ACTH was initiated by low corticosterone levels in the blood of rats.

Cheng and Sayers (1950) observed that administration of a mineralocorticoid, desoxycorticosterone, increased the concentration of ACTH in the adenohypophysis of the intact rat. McIntosh and Holmes (1951) suggested that cortisone inhibits adrenal corticoid synthesis and release by making the adrenal cortex refractive to ACTH. But Fortier *et al.* (1951) found that cortisone injection failed to prevent or decrease adrenal ascorbic acid depletion indicating that cortisone administration must act on the pituitary to inhibit ACTH release. Farrell and Laqueur (1955) showed that cortisone acts on the pituitary. Dogs were given subcutaneous injections of 100 mg of cortisone daily, and it was found that after 14 days of this treatment, pituitary ACTH was depressed 54%. Hodges and Sadow (1967) found similar results in the rat.

The controversy whether corticoids act on the pituitary or hypothalamus was partially resolved by the work of Davidson *et al.* (1963) and Stark *et al.* (1967). The former workers noted that hydrocortisone placed in culture with pituitary glands did not cause a decrease in ACTH secretion, but when hypothalamic extracts were included ACTH secretion was inhibited. Stark and his colleagues observed that only hydrocortisone implants in the median eminence inhibited ACTH secretion. On the basis of electron microscopic studies by Herlant (1964) on rats and by Yamada and Yamashita (1967) on mice, it was concluded that hydrocortisone caused inhibition of the corticotropes.

Ultrastructural Changes in Corticotropes After Administration of Hydrocortisone

Hydrocortisone treatment for 1–2 days caused little change in the structure of the anterior pituitary gland. After treatment for 5–7 days part of the corticotropes were well granulated (Fig. 5.14) but the Golgi appeared to be inactive. The endoplasmic reticulum consisted of dilated cisternae in some cells and poorly organized fragments in other cells. Some ribosomes were free in the cytoplasm which may indicate a declin-

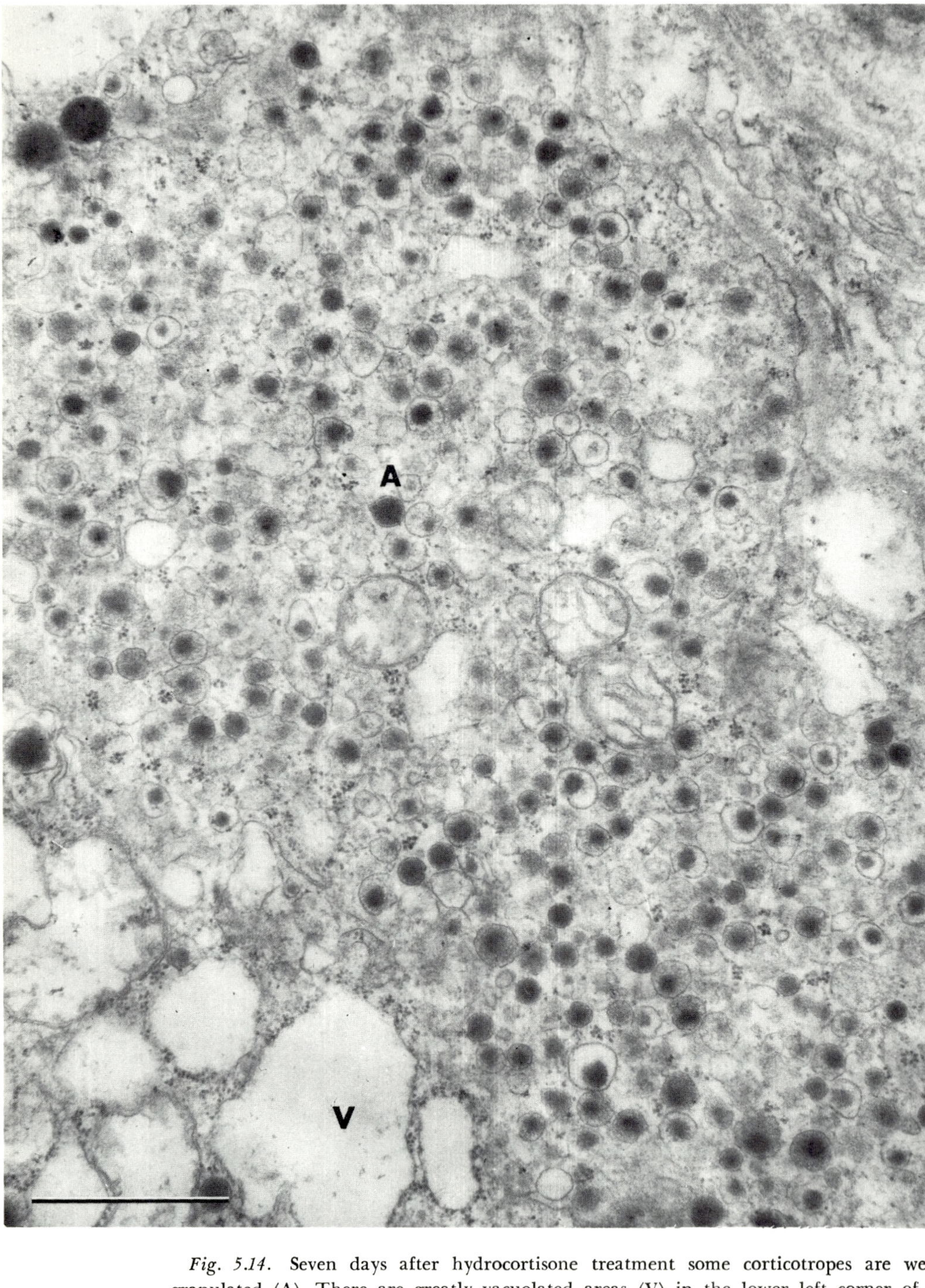

Fig. 5.14. Seven days after hydrocortisone treatment some corticotropes are well granulated (A). There are greatly vacuolated areas (V) in the lower left corner of a somatotrope. × 30,000. (Veronal acetate buffered osmic acid)

ing function of these cells. Mitochondria were present as round or short rods, and there were a few lysosomes. Other cell types were not changed.

Hydrocortisone treatment for 14 days resulted in the so-called Crook's changes in the pituitary; that is, the cytoplasm was greatly vacuolated (Fig. 5.15). Corticotropes were not observed at this time and it was difficult to identify the other cell types because their cell membranes were indistinguishable. Hydrocortisone injections for 30–60 days caused greater pathological changes than those caused by other treatments. The mitochondria were round and swollen. Many vacuoles, fat droplets, and an absence of a Golgi complex characterized many of the cell types. There was also an infiltration of connective tissue into areas around the capillaries. Hydrocortisone selectively affected the fine structure of the corticotropes, but with longer administration the general cytology of the gland was also greatly changed.

Discussion of Ultrastructural Findings

Hydrocortisone administration of 1–5 days had little effect on the rat pituitary. The general cytology of the gland was normal except that a few of the corticotropes were hypertrophied. This may indicate that more ACTH was stored than secreted. However, hydrocortisone administered for 7–60 days resulted in atrophy or degeneration of corticotropes and other hypophyseal cells. Fourteen days after treatment the pituitary was devoid of corticotropes and the general ultrastructure was abnormal. The cytoplasm was degranulated and vacuolated; the cells lacked membranes and fatty inclusions were evident. Schelin (1962) and Selye (1950) reported pituitary tumors resulting from long-term administration of stressors, including large doses of cortisone. Thornton (1956) reported that if cortisone treatment was stopped these pathological changes disappeared. The present findings are in agreement with the results of Herlant (1964) who also observed abnormal changes in pituitary cells, especially corticotropes, after hydrocortisone treatment.

EXOGENOUS ACTH

After chronic administration of ACTH, Evans *et al.* (1943) reported hypertrophy of the adrenal cortex in immature rats. When ACTH was administered to rats for 30 days, Koneff (1944) found a decrease in the weight of the rat pituitary gland and a diminution in basophils but no effect on the acidophils of the gland. Thornton (1956) described similar findings and hyalinization of basophils.

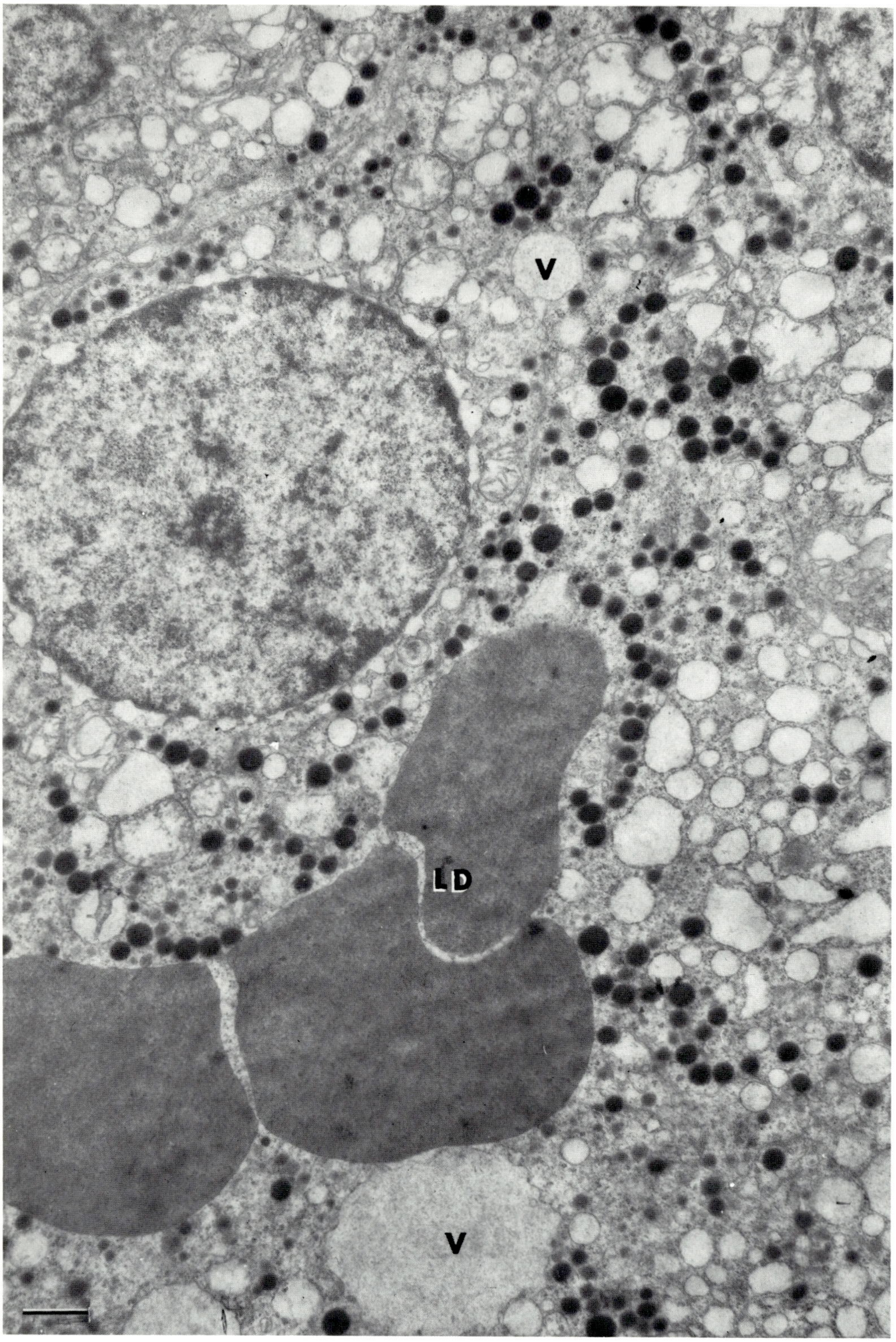

Fig. 5.15. Lipid droplets (LD) and a vacuolated cytoplasm (V) are indicative of the pituitary after 30 days of hydrocortisone treatment. × 10,000. (Veronal acetate buffered osmic acid)

Kitay *et al.* (1958) have shown that daily injection of 4 USP units of ACTH for 7 days caused an increase in ACTH and concluded that high blood levels might inhibit release of endogenous ACTH. Kitay (1959) and Kitay *et al.* (1959) in additional studies noted that ACTH given to intact rats brought about a 95% rise in pituitary ACTH, but when given to adrenalectomized animals it caused a doubling in pituitary ACTH over the control levels. This exogenous ACTH also inhibited the usual hypersecretion of endogenous ACTH following application of scalding stress.

There is evidence that exogenously administered ACTH prevents or inhibits the release of this hormone from the pituitary gland.

Ultrastructure of Corticotropes After ACTH Treatment

The corticotropes hypertrophied after ACTH treatment. Many were well granulated and the granules were denser than in control pituitaries (Fig. 5.16). The mitochondria were often normal in outline, but in a few cells they were swollen. The number of lysosomes was approximately equal to that found in intact tissue. Golgi complexes were dilated in many cells and the endoplasmic reticulum was irregular and usually hypertrophied.

Exogenous corticotropin greatly stimulated the corticotropes. It probably acted on the adrenal cortex which allowed the corticotropes of the pituitary to store their hormone in granules rather than secrete it.

Discussion of ACTH Administration Effect on Corticotrope Ultrastructure

After short-term administration of ACTH to rats, the pituitary appeared similar to the intact controls. However, more granules were present in the corticotropes and there was an increase in free and attached ribosomes. It appeared that the endogenous ACTH was stored and that the exogenous ACTH served to maintain the adrenal cortex. These findings are in agreement with the data of Kitay (1959) and Kitay *et al.* (1958, 1959) who reported an increase in ACTH after exogenous ACTH administration. Sussman *et al.* (1965) also found in human subjects that exogenous ACTH inhibited release of ACTH. Similarly, Krulich and McCann (1966) concluded that somatotropin administered to intact rats for 6 days caused an increase of this hormone in the pituitary. However, they noted that long-term STH treatment resulted in a decrease in pituitary somatotropin. In this connection, Gemzell and Heijkenskjold (1957) found that long-term ACTH treatment reduced the pituitary content of this hormone.

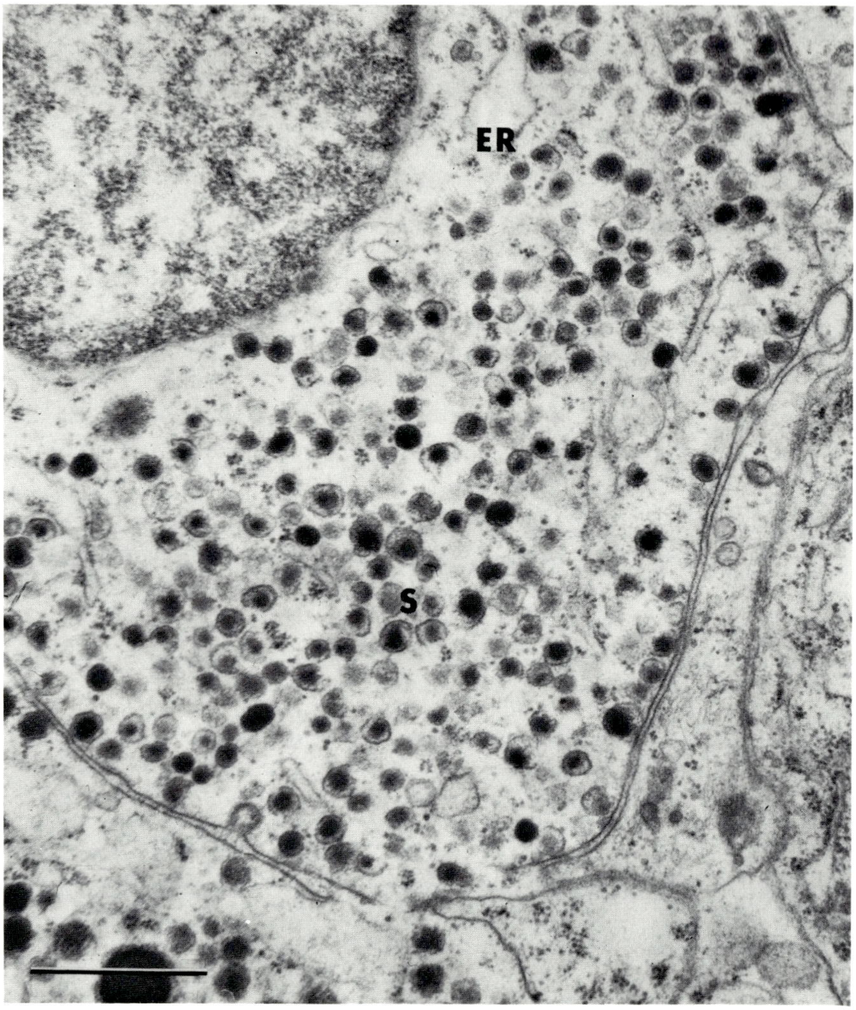

Fig. 5.16. A corticotrope after ACTH treatment. An abundance of granules (S) and several dilated cisternae of endoplasmic reticulum (ER) are shown. × 23,450. (Veronal acetate buffered osmic acid)

α-ETHYLTRYPTAMINE

Tullner and Hertz (1964) observed hypotension following the administration of 3-(2-aminobutyl)-indole acetate (monase, α-ethyltryptamine). They observed that the intravenous injection of 5 mg/kg α-ethyltryptamine to dogs, under phenobarbital anesthesia, resulted in a marked decrease in adrenal steroid output which returned to normal 2 hours after administration. If exogenous ACTH was injected after α-ethyltryptamine, there was no depression. These data suggest that α-ethyltryptamine suppresses the formation or release of ACTH.

Lorenzen *et al.* (1965), Lorenzen and Ganong (1967), and Ganong *et al.* (1965) have reported the mode of action of ethyltryptamine. Like Tullner and Hertz (1964), they also found that this compound depresses endogenous ACTH secretion. They concluded that it did not act on the adrenal cortex since the adrenal response to exogenous ACTH was not changed. Stimulation of the ventral hypothalamus, particularly the median eminence, produced an increase in 17-hydroxycorticoids which was not blocked by α-ethyltryptamine. It was suggested that α-ethyltryptamine inhibited ACTH secretion by its action on the hypothalamus and that it may prevent release of the corticotropin-releasing factor (CRF).

In further work the Ganong group (1967) observed in stressed dogs that when a rise in blood pressure, caused by α-ethyltryptamine, is prevented by hemorrhage, an inhibition of ACTH does not occur. The results suggest that it is the rise in blood pressure produced by α-ethyltryptamine that causes the inhibition of ACTH secretion.

It appears that α-ethyltryptamine prevents the synthesis and/or release of ACTH by blocking CRF release at the hypothalamic level.

Ultrastructure of Corticotropes After Administration of α-Ethyltryptamine

There appeared to be more corticotropes and these were hypertrophied due to α-ethyltryptamine treatment. They often had more granules which were larger and crowded so close together that the Golgi and endoplasmic reticulum could not be seen. Usually the ACTH granules were more uniformly electron dense than in control animals (Fig. 5.17). Mitochondria were slightly swollen in many corticotropes and a few lysosomes and cilia were seen in some ACTH cells.

It is suggested that α-ethlytryptamine prevents the release but apparently not the synthesis of ACTH since these cells were well granulated.

Discussion of Results From αEthyltryptamine Treatment

There was an increase in ACTH cells after treatment with α-ethyltryptamine and there were more granules in the majority of the corticotropes than in pituitary glands from intact animals. It has been reported

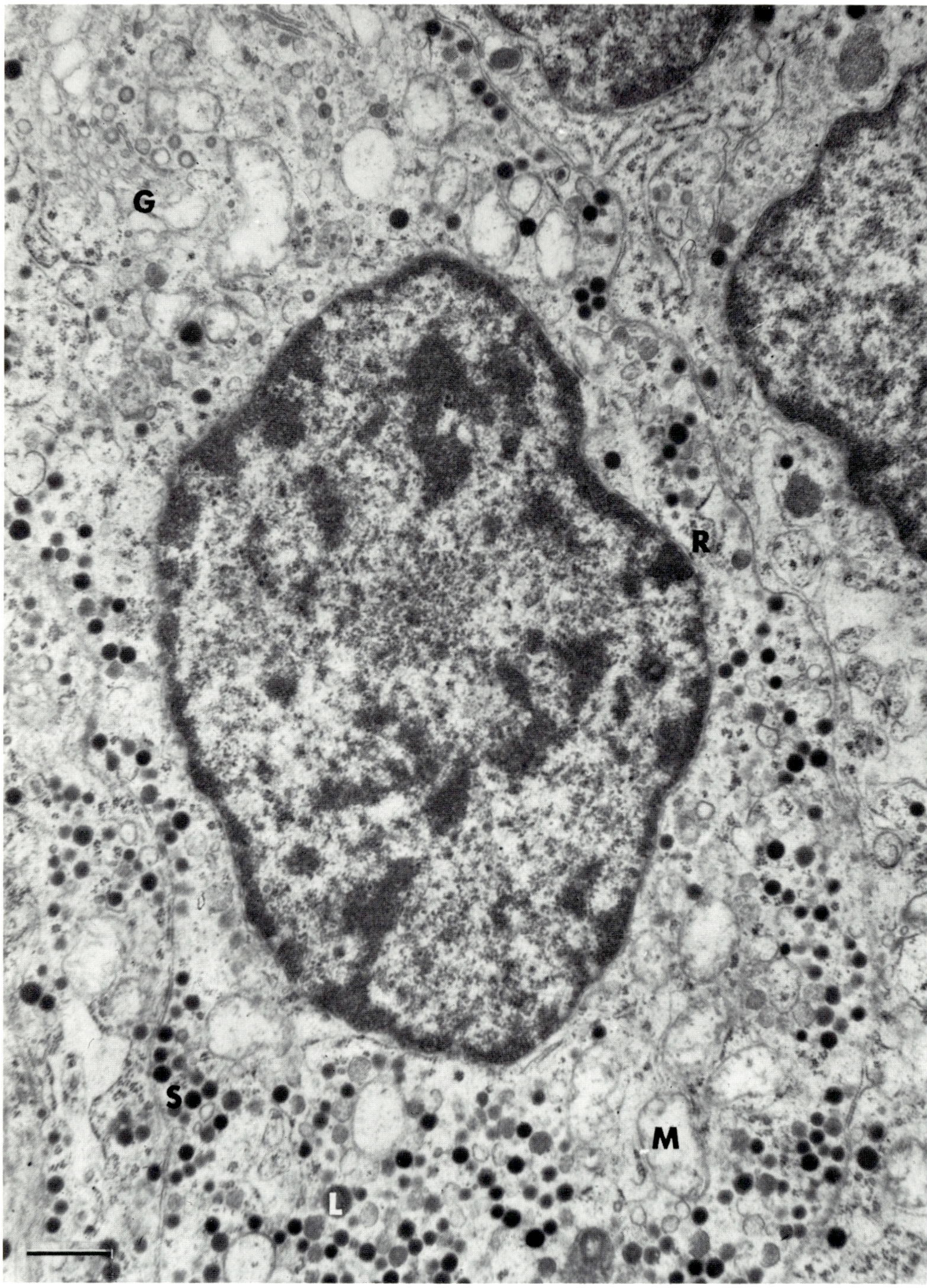

Fig. 5.17. A corticotrope after αethyltryptamine treatment. There is an abundance of granules (S) of high density and increased diameter. A stimulated Golgi complex (G) and hypertrophied mitochondria with obliterated cristae (M), lysosomes (L), and free ribosomes (R) are shown in this micrograph. $\times$ 12,500. (Veronal acetate buffered osmic acid)

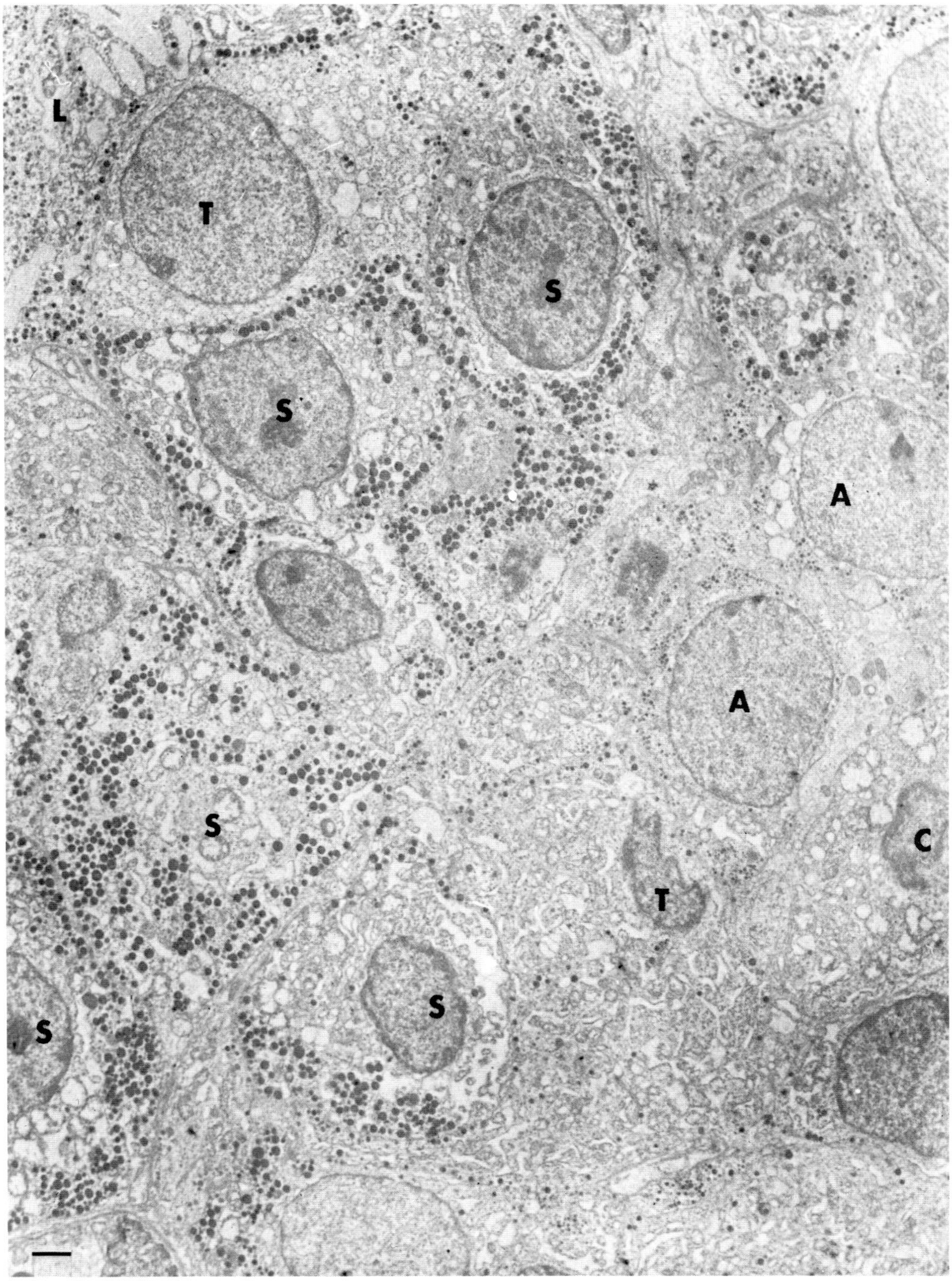

Fig. 5.18. A section of male rat anterior pituitary gland after 7 days of adrenalectomy and castration. Somatropes (S), a LH cell (L), thyrotropes (T), corticotropes (A), and a chromophobe (C) are shown. Generally, the cells are more vacuolated and mitochondria are swollen. × 4950. (Veronal acetate buffered osmic acid)

by Ganong *et al.* (1965, 1967) that α-ethyltryptamine blocks the release of CRF from the hypothalamus, but in the present studies other pituitary cells in addition to the corticotropes were also well granulated.

CORTICOTROPES AFTER ADRENALECTOMY AND CASTRATION

After adrenalectomy and castration corticotropes appeared to be like those after adrenalectomy alone (Fig. 5.18). The mitochondria were swollen with interrupted cristae and the granulation was often greater than in the corticotropes of intact rats. The ACTH cells were not as large or irregular in shape as in the corticotropes of adrenalectomized rats. D'Angelo and Hughes (1967) and Coyne and Kitay (1968, 1969) reported a decrease in blood ACTH after castration indicating that more ACTH was stored. Richard (1965) showed that estrogen either promoted release of CRF from the hypothalmus or it acted directly on the pituitary to cause release of ACTH. Accordingly, castration and the lack of estrogen would prevent release of high levels of CRF and more corticotropin would be stored than released.

GENERAL DISCUSSION OF CORTICOTROPES

Location of Corticotropes

According to our fine structure studies of rat pituitaries, intact, adrenalectomized, and treated rats, corticotropes predominate in the posterior-medial region (Costoff, 1969). Smelser (1944) and Rochefort and Saffran (1957) an abundance of ACTH in the medial regions of pituitaries from various species.

Although each pituitary cell type may be located throughout the gland, each tends to be more abundant in certain parts of the gland. Adams *et al.* (1964) transected the pituitary stalks of various species of animals and found that groups of portal vessels passing through the stalk supply circumscribed regions in the anterior pituitary. They suggested that the neurovascular link between the hypothalamic nuclei and the pars distalis included many pathways each with circumscribed regions of origin as well as termination. They concluded that individual groups of similar cells of the anterior lobe receive specific neurohumors produced by particular groups of hypothalamic nerve cells, i.e., the vascular supply to the corticotropes transports only CRF and not other releasing factors.

Corticotropes after Adrenalectomy

Our findings showed that ACTH cells were the only cell type that changed after adrenalectomy. They were somewhat degranulated at 1 or 2 days, but after 5 days there were more corticotropes and increased granulation. Since few mitotic figures were observed, the question arises as to how the ACTH cells increased in number.

Hunt (1942, 1943, 1951) reported that in pituitaries from mature rats only 1 out of 1000 cells contained a mitotic figure whereas this probability was increased in female rats in estrus to 20 per 1000 cells. Dhom and Stöcker (1964), Crane *et al.* (1965) from studies in intact rats, and Kraicer and Cheng (1968) and Kraicer *et al.* (1967) in adrenalectomized rats, concluded that there were few mitotic figures in the pituitaries of these animals. It appears, therefore, that hyperplasia of ACTH cells does not occur extensively by mitosis.

There is the possibility that chromophobes give rise to corticotropes. Severinghaus (1937) proposed that chromophobe mother-stem cells develop into acidophic and basophilic chromophobe cells, which in turn change into the respective chromophils. The concept of Collin (1933) was similar to that of Severinghaus except that he postulated that chromophils could dedifferentiate into chromophobes. Based on autoradiographic studies Siperstein (1963) and Knutson (1966b) it was observed that as the length of time following adrenalectomy increased more small chromophobes differentiated into larger chromophobes, and subsequently into adrenalectomy cells. Histochemical studies by Totsuka (1967) also support the hypothesis that chromophobes give rise to ACTH cells. This hypothesis can be extended further by postulating, as Collin (1933) did, that degranulated somatotropes 4 days after adrenalectomy may degenerate into chromophobes which in turn may differentiate into ACTH cells.

Pituitary Corticotropin After Adrenalectomy

In our studies, the corticotropes were somewhat degranulated 1 or 2 days after adrenalectomy. After 4 or 5 days there was an increase in both the number and granulation of ACTH cells. Seven days post adrenalectomy there were slightly fewer ACTH cells, with fewer granules. These findings agree with the results of Kurosumi and Kobayashi (1966), except they found degenerated corticotropes 7 days after adrenalectomy. In the present work there was degranulation of some corticotropes at 7 days following adrenalectomy but others were well granulated at 14, 30, and 60 days. Yamada and Yamashita (1967) observed well-granulated ACTH cells in mouse glands 14 days after adrenalectomy, which agrees with our findings.

Corticotropin synthesis in the rat pituitary does not decrease or stop 7 days after adrenalectomy as Kurosumi and Kobayashi suggest (1966). Sayers and Cheng (1949) found that the ACTH content of the pituitary was reduced 80%, 24 hours after adrenalectomy. This agrees with the present findings in that there was degranulation of corticotropes 1 day after adrenalectomy. The bioassay results of other workers (Gemzell *et al.*, 1951; Fortier,, 1959; Syndor and Sayers, 1954) found a 40% drop in pituitary ACTH after the first day of adrenalectomy but a return to normal concentration after 4 days. After 7 days these workers reported pituitary ACTH levels to be 50% above normal with a peak at 32 days after adrenalectomy. ACTH levels were normal or elevated in animals for several months after adrenalectomy. These data support our findings. Also Cox *et al.* (1958) concluded that it requires 2 weeks for the adrenalectomized rat pituitary to secrete and maintain a high output of ACTH. Hodges and Vernikos-Danellis (1962) also observed that long-term adrenalectomized animals responded more to stress than did the controls. Later, Vernikos-Danellis (1965) reported CRF content in the median eminence increased 5 days post adrenalectomy with a peak at 20 days. These results show a parallel between CRF and ACTH levels after adrenalectomy.

The physiological findings and bioassay data support the research described here, namely, that ACTH synthesis and secretion continues for a long time after adrenalectomy. However, there appear to be too few corticotropes to account for the doubling of ACTH in the pituitary and 30-fold increase of it in the blood after long-term adrenalectomy (Gemzell *et al.*, 1951). We agree with Siperstein (1963) in that although there is an increase in corticotropes 5 days after adrenalectomy this increase in cells would not account for the great output of this hormone for long periods after adrenalectomy. Siperstein estimated that 1.5% of all the pituitary cells are ACTH cells, we have estimated as many as 4% are corticotropes. But Gemzell *et al.* (1951) report that ACTH accounted for 13% of the hormone content of pituitary glands from rats adrenalectomized for 1 week. There must be another source of ACTH, in addition to the corticotropes described above that is operative after long-term adrenalectomy. A question arises regarding the source of this ACTH.

If there are not sufficient corticotropes and granules in these cells, perhaps ACTH may remain in the cisternae of the endoplasmic reticulum and be released directly from the cell without granules being formed. Palay (1958) postulated that the cisternae of the endoplasmic reticulum may have openings to the outside of the cell so the hormone could be released without condensation into granules. In this connection Tixier-Vidal and Picart (1967) postulated that LTH is released directly into the capillaries without granule formation. Weiss (1965) also suggested that if the demand is great enough the hormone in the platyfish pituitary may be released without storage in granules.

Another hypothesis that may account for the increased ACTH content in the pituitary after long-term adrenalectomy was reported by Farquhar (1957). She suggested that ACTH is produced by a type of chromophobe (follicle cells) and that ACTH might be stored in the follicle (see Chap. 8). Since a degranulated corticotrope is similar to a follicle cell the two cells may be related. Kagayama (1965) showed in dog pituitary glands that the volume of colloid found in the follicle varied with physiological conditions. This is in agreement with the present findings. Since these follicle cells have a surface composed of microvilli and cilia, it is suggestive that material is being stored in and absorbed from the follicle.

A third hypothesis to account for the large amount of ACTH found in the pituitary after long-term adrenalectomy is that of Kastin *et al.* (1967). These investigators reported that MSH activity and granulation of MSH cells in the pars intermedia of rats is depressed following stress. This inhibition of MSH parallels the depression of ACTH following stress. Kobayashi (1965) proposed that a secretory product of the pars intermedia, presumably MSH, influences the adrenal cortex. Recently, Gosbee and Kraicer (1970) described the results of autoradiographic studies of the rat pituitary gland 1 month after adrenalectomy. They did not find an increase in label in the cells of the pars distalis but there was an increase in the label in the cells of the pars intermedia. They suggested that after adrenalectomy, cells of the pars intermedia are active and may be implicated in corticotropin synthesis and release.

The evidence for the increase in pituitary ACTH after long-term adrenalectomy cannot be wholly accounted for by the amount of hormone associated with granules present in the cell identified as a corticotrope. It appears that the hormone is released as soon as it is formed, is stored in a follicle prior to release, or perhaps MSH cells or another type of cell of the pars intermedia produce an ACTH-like hormone.

BIOLOGICAL ACTIVITIES OF PITUITARY GRANULE FRACTIONS

Corticotropin activity of the various granule fractions were assayed by the Saffran and Schally (1955) *in vitro* method. Over 53% of the

ACTH activity was found in the D1HSP pellet (Fig. 5.19). The SA soluble zone contained 19.1% of the activity. This indicates the relative instability of these granules because many were evidently solubilized. About 10% of the activity was found in the FD2HSP pellet, but only 2.4% in the FD2LSP pellet. The activities of the FD2 low- and high-speed pellets may be due to not removing the entire D1 zone from the gradient. Granules in the D1HSP pellet have a mean diameter of 89 mμ (Costoff and McShan, 1969). Perdue and McShan (1966) found ACTH, TSH, and the gonadotropins associated with a fraction of granules that measured 140–200 mμ in diameter. Diameters of granules in corticotropes did not change after adrenalectomy or treatment with blocking agents. After α-ethyltryptamine administration, granules in ACTH cells were increased up to 30 mμ in diameter.

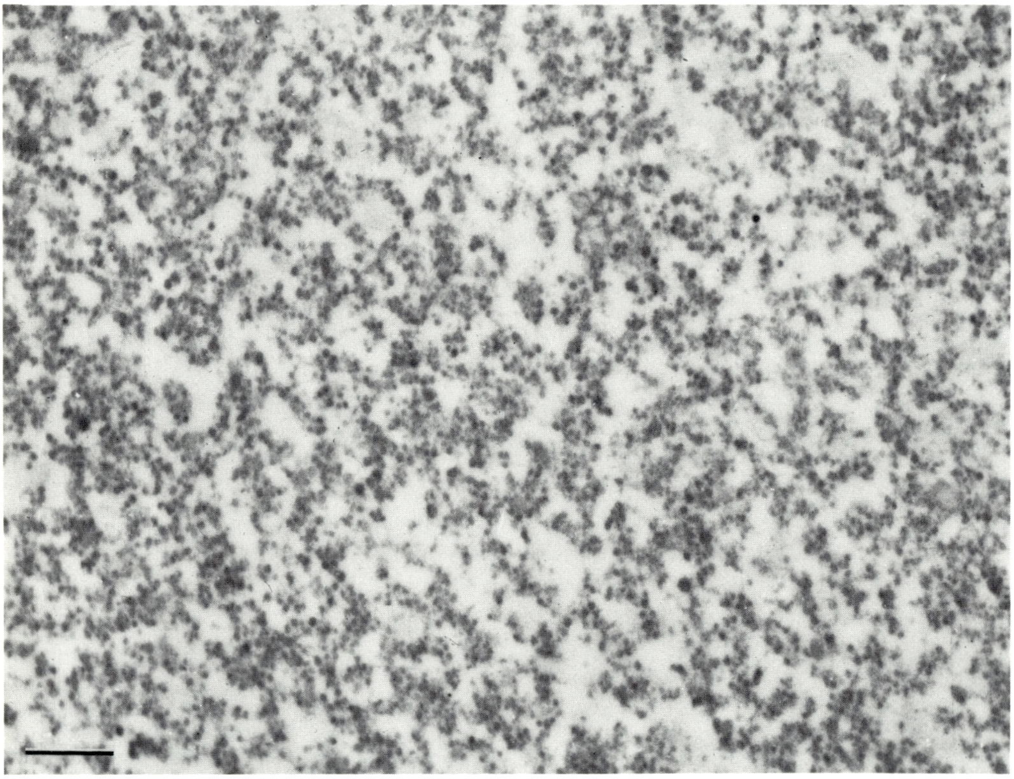

Fig. 5.19. A section of the DIHSP pellet. ACTH activity was associated with granules of this preparation. $\times$ 12,000. (Veronal acetate buffered osmic acid)

Summary

From the histochemical standpoint the identity of the corticotrope may have been established but there is still some controversy over the type of cell producing ACTH. Herlant (1964) reported that corticotropes could be identified with the tetrachrome staining method and believes that corticotropes are acidophils. Data from immunofluorescent studies reported by Hess *et al.* (1968) and Baker (1970); and peroxidase labeled antibody findings of Baker *et al.* (1970) indicate that the basophil which stains weakly with PAS is the source of ACTH. Since ACTH is not a glycoprotein like TSH, FSH, and LH, the possibility exists that a polysaccharide might somehow become associated with the hormone or the ACTH granule.

It seems plausible that the large chromophobe-type cell that Siperstein (1963) has shown to be active after adrenalectomy is the same cell that has been identified by electron microscopy as the source of ACTH. Rennels (1964) shows micrographs of well-granulated corticotropes in glands from rats in which stress was produced by scalding. Siperstein and Allison (1965), Kurosumi and Kobayashi (1966), Kurosumi and Oota (1966), and Siperstein and Miller (1970) have described criteria for corticotropes. The ultrastructural studies of Siperstein and Miller (1970), however, show some corticotropes after cortisol treatment that have granules over 300 mμ in diameter. We have not seen granules of that size or oblong shape in cells which we designate as corticotropes. Nakayama *et al.* (1969) did adrenal cortical regeneration studies and described their criteria for identification of ACTH cells. They used female rats that were killed at unknown times during the estrous cycle. Some of the cells they identified as corticotropes appear to be thyrotropes. Thyrotropes are larger and better granulated in female rat pituitary glands. However, Nakane (1970) reports that it is difficult to distinguish between TSH and ACTH cells in his peroxidase-labeled antibody studies. Kurosumi and Oota (1966) however, have shown in rats that corticotropes differ from thyrotropes and gonadotropes in different physiological states.

REFERENCES

Adams, J. H., Daniel, P. M., and Prichard, M. M. (1964). Distribution of hypophyseal portal blood in the anterior lobe of the pituitary gland. *Endocrinology* **75**, 120.

Addison, T. (1855). "On the Constitutional and Local Effects of Disease of the Suprarenal Capsule." D. Highley, London.

Allen, M. J. (1950). Preparation of some substituted glycols by electrolytic reduction. *J. Org. Chem.* **15**, 435.

Allen, M. J., and Corwin, A. H. (1951). Chemical studies on pinacols obtained from P-aminoacetophenone and P-aminopropiophenone. *J. Amer. Chem. Soc.* **72**, 117.

Arimura, A., Saito, T., and Schally, A. V. (1967). Assays for corticotropin-releasing factor using rats treated with morphine, chlorpromazine, dexamethasone, and Nembutal. *Endocrinology* **81**, 235.

Arimura, A., Bowers, C. Y., Schally, A. V., Saito, T. M., and Miller, M. C. (1969). Effect of CRF, dexamethasone and actinomycin D on release of ACTH from rat pituitaries *in vivo* and *in vitro*. *Endocrinology* **85**, 300.

Baker, B. L. (1970). Studies on hormone localization with emphasis on the hypophysis. *J. Histochem. Cytochem.* **18**, 1.

Baker, B. L., Pek, S., Midgley, J. R., Jr., and Gersten, B. E. (1970). Identification of the corticotropin cell in the rat hypophysis with peroxidase-labeled antibody. *Anat. Rec.* **166**, 557.

Bencze, W. L., and Allen, M. J. (1957). Structure of amphenone B and related amphenones. *J. Org. Chem.* **22**, 352.

Bernhard, W. (1958). Electron microscopy of tumor cells and tumor viruses. *Cancer Res.* **18**, 491.

Brozman, M. (1967). Histochemical localization of ACTH and TSH in the human hypophysis. *Acta Histochem.* **26**, 261.

Burt, A. S., and Verlardo, J. (1954). Cytology of human adenohypophysis as related to bioassays for tropic hormones. *J. Clin. Endocrinol. Metab.* **14**, 979.

Burt, A. S., Landing, B. H., and Sommers, S. C. (1954). Amphophil tumors of the hypophysis induced in mice by I^{131}. *Cancer Res.* **14**, 497.

Chart, J. J., Sheppard, H., Allen, M. J., Bencze, W. L., and Gaunt, R. (1958). New amphenone analogs as adrenocortical inhibitors. *Experientia* **14**, 151.

Cheng, C., and Sayers, G. (1950). Desoxycorticosterone acetate and adenohypophyseal content of adrenocorticotrophic hormone. *Proc. Soc. Exp. Biol. Med.* **74**, 674.

Cheng, S. C., and Carballeira, A. (1969). Effects of Metopirone on hydroxylation of cholesterol by rat adrenal mitochondria. *Endocrine Soc. Meet.* p. 75 (abstr.).

Cleveland, R., and Wolfe, J. M. (1933). Cyclic histological variations in the anterior hypophysis of the sow. *Amer. J. Anat.* **53**, 191.

Collin, R. (1933). "L'Hypophyse Traveau Origineaux et Étude." George Thomas, Nancy.

Costoff, A. (1969). Isolation of secretory granules and the study of the fine structure of the male rat anterior pituitary glands. Ph.D. Thesis, University of Wisconsin, Madison.

Costoff, A., and McShan, W. H. (1969). Isolation and biological properties of secretory granules from rat anterior pituitary glands. *J. Cell Biol.* **43**, 564.

Cox, G. S., Hodges, J. R., and Vernikos, J. (1958). The effect of adrenalectomy on the circulating level of ACTH in the rat. *J. Endocrinol.* **17**, 17.

Coyne, M. D., and Kitay, J. I. (1968). Effect of ovariectomy on pituitary ACTH secretion in rats. *Fed. Proc., Fed. Amer. Soc. Exp. Biol.* **27**, 434.

Coyne, M. D., and Kitay, J. I. (1969). Effect of ovariectomy on pituitary secretion of ACTH. *Endocrinology* **85**, 1097.

Crane, W. A., Dutta, L. P., and Ingle, D. J. (1965). Cell proliferation in the rat pituitary. *Proc. Soc. Exp. Biol. Med.* **119**, 167.

Crooke, A. C., and Russell, D. (1935). The pituitary gland in Addison's disease. *J. Pathol. Bacteriol.* **40**, 255.

Cushing, H. (1932). The basophil adenomas and their clinical manifestations. *Bull. Johns Hopkins Hosp.* **50**, **137**.

D'Angelo, S. A., and Hughes, J. S. (1967). Chronic hypothalamic lesions and anterior pituitary secretion: Effects of ovariectomy on TSH and ACTH. *Neuroendocrinology* **2**, 263.

D'Angelo, S. A., Gordon, A. S., and Charipper, H. A. (1948). The effect of inanition on the anterior pituitary-adrenocortical interrelationships in the guinea pig. *Endocrinology* **42**, 399.

Davidson, J. M., and Feldman, S. (1967). Effects of extra hypothalamic dexamethasone implants on the pituitary-adrenal system. *Acta Endocrinol. (Copenhagen)* **55**, 240.

Davidson, J. M., Chowers, I., and Feldman, S. (1963). ACTH secretion in rats with intracerebral hydrocortisone implants. *Fed. Proc., Fed. Amer. Soc. Exp. Biol.* **22**, 571 (abstr.).

Descourt, M., and Herlant, M. (1963). Etude d'un adenoma hypophysaire avec syndrone de cushing favorisant l'identification de cellules cortiocotrops chez l'hormone. *Ann. Endocrionol.* **24**, 497.

de Duve, C., and Wattiaux, R. (1966). Functions of lysosomes. *Annu. Rev. Physiol.* **28**, 435.

De Wied, D. (1962). The site of the blocking actions of dexamethasone on stress induced pituitary ACTH release. *Acta Endocrinol. Copenhagen* **67**, Suppl., 75.

Dhom, G., and Stöcker, E. (1964). Autoradiographische studien über die DNS-synthese im hypophysen vorderlappen der ratte. *Experientia* **20**, 384.

Dhom, G., Burkert, F., and Tietze, H. V. (1962). Thyroidectomy cells in the anterior hypophysis of rats after adrenal gland blockade. On the problem of the site of formation of ACTH. *Z. Zellforsch. Mikrosk. Anat.* **57**, 679.

Dominguez, O. V., and Samuels, L. T. (1963). Mechanism of inhibition of adrenal steroid 11β-hydroxylase by metopyrapone. *Endocrinology* **73**, 304.

Dubois, M. P., and Herlant, M. (1968). Cytological characteristics of gonadotroph, thyrotroph, somatotrophs and prolactin cells in the anterior lobe of cattle pituitary. *Ann. Biol. Anim., Biochim., Biophys.* **8**, 5.

Evans, H. M., Simpson, M. E., and Li, C. H. (1943). Inhibiting effect of adrencorticotropic hormone on the growth of male rats. *Endocrinology* **33**, 237.

Ezrin, C., Wilson, W. D., Dawson, J. W., and Hill, F. M. (1956). The cytology of the adenohypophysis in various disorders of the adrenal cortex. *J. Clin. Endocrinol. Metab.* **16**, 937.

Farmer, T. A. (1961). The plasma 17-Hydroxycorticosteriod response to ACTH, Metopirone and lipopolysaccharide pyrogen in the human. *J. Clin. Endocrinol. Metab.* **21**, 433.

Farquhar, M. G. (1957). "Corticotrophs" of rat adenohypophysis as revealed by electron microscopy. *Anat. Rec.* **127**, 291.

Farrell, G. L., and Laqueur, G. (1955). Reduction of pituitary content of ACTH by cortisone. *Endocrinology* **56**, 471.

Field, E. J. (1958). The effects of adrenalectomy on the anterior lobe of the pituitary gland of the rat. *J. Anat.* **92**, 137.

Finerty, J. C., and Briesno-Castrejon, B. (1949). Quantitative studies of cell types on the rat anterior pituitary following unilateral adrenalectomy. Endocrinology **44**, 293.

Finerty, J. C., Hess, M., and Binhammer, E. (1952). Pituitary cytological manifestations of heightened adrenocortocotropic activity. *Anat. Rec.* **114**, **15**.

Fortier, C. (1959). Pituitary ACTH and plasma free corticosteroids following bilateral adrenalectomy in the rat. *Proc. Soc. Exp. Biol. Med.* **100**, 13.

Fortier, C., Yrarrazaval, S., and Selye, H. (1951). Limitations of ACTH regulating effect of cortisone. *Amer. J. Physiol.* **165**, 467.

Furth, J., Gadsen, E. L., and Upton, A. C. (1953). ACTH secreting transplantable pituitary tumors. *Pro. Soc. Exp. Biol. Med.* **84**, 253.

Ganong, W. F., Wise, B. L., Shackleford, R., Boryczka, A. T., and Zipf, B. (1965). Site at which α-ethyltryptamine acts to inhibit the secretion of ACTH. *Endocrinology* **76**, 526.

Ganong, W. F., Borycza, A. T., Lorenzen, L. C., and Egge, A. S. (1967). Lack of effect of α-ethyltryptamine on ACTH secretion when blood pressure is held constant. *Proc. Soc. Exp. Biol. Med.* **124**, 558.

Garweg, G., Joussen, F., and Kinsky, I. (1969). Autoradiographische untersuchungen über die neurosekretorischen sustanzen im hypothalamus der mans nach Metopiron. *Z. Zellforsch. Mikrosk. Anat.* **101**, 510.

Gemzell, C. A., and Heijkenskjold, F. (1957). Effect of corticotrophin on the content of corticotrophin in the pituitary gland of adrenalectomized rats. *Acta Endocrinol. (Copenhagen)* **24**, 249.

Gemzell, C. A., VanDyke, D. C., Tobias, C. A., and Evans, H. M. (1951). Increase in the formation and secretion of ACTH following adrenalectomy. *Endocrinology* **49**, 325.

Gold, E. M., Bigler, R., Newman, S., Angers, M. and DiRaimondo, V. (1959). Studies of pituitary adrenocortical relationships with an 11β-hydroxylase inhibitor of the adrenal cortex. *Clin. Res.* **7**, 64.

Gold, E. M., Kent, J. R., and Forsham, P. H. (1961). The clinical use of methopyrapone in pituitary and adrenocortical disorders. *Ann. Intern. Med.* **54**, 175.

Gosbee, J. L., and Kraicer, J. (1970). Autoradiographic study of cell replication in pars distalis with altered ACTH secretion. *Fed. Proc. Fed. Amer. Soc. Exp. Biol. Med.* **29**, 311 (abstr.).

Griffiths, K. (1963). Inhibition of '19 hydroxylase' activity in the golden hamster adrenal by SU4885. *J. Endocrinol.* **26**, 445.

Grollman, A., and Firor, W. M. (1935). The role of the hypophysis in experimental chronic adrenal insufficiency. *Amer. J. Physiol.* **112**, 310.

Halmi, N. S., and Bogdanove, E. M. (1951). Effect of thyroidectomy on ACTH content of the rat adenohypophysis. *Proc. Soc. Exp. Biol. Med.* **77**, 518.

Heinbecker, P., and Rolf, D. (1944). Hypophyseal eosinophil cell and insulin sensitivity. *Amer. J. Physiol.* **141**, 566.

Herlant, M. (1953). La corticotropine est bein élaborée par les cellules acidophiles de l'hypophyse. *Ann. Endocrinol.* **14**, 64.

Herlant, M. (1964). The cells of the adenohypophysis and their functional significance. *Int. Rev. Cytol.* **17**, 299.

Hertz, R., and Tullner, W. W. (1950). The effect of amphenone "B" on thyroid, adrenals and genital tract of the female rat. *Proc. Soc. Exp. Biol. Med.* **75**, 627.

Hertz, R., Tullner, W. W., Schricker, J. A., Dhyse, F. G., and Hallman, L. F. (1955). Studies on amphenone and related compounds. *Recent Progr. Horm. Res.* **11**, 119.

Hess, M., and Hall, C. E. (1951). The effects of desoxycorticosterone acetate and hypertension upon pituitary gland cytology in the rat. *Anat. Rec.* **109**, 304.

Hess, R., Barratt, D., and Gelzer, J. (1968). Immunofluorescent localization of β-corticotropin in the rat pituitary. *Experientia* **24**, 584.

Hodges, J. R., and Sadow, J. (1967). Impairment of pituitary adrenocorticotrophic function by corticosterone in the blood. *Brit. J. Pharmacol. Chemother* **30,** 385.

Hodges, J. R., and Vernikos-Danellis, J. (1962). Pituitary and blood corticotropin changes in adrenalectomized rats maintained on physiological doses of corticosteroids. *Acta Endocrinol. (Copenhagen)* **39,** 79.

Hogness, J. R., Lee, N. D., and Williams, R. H. (1953). Effects of amphenone "B" on thyroid and adrenal functions. *Endocrinology* **52,** 378.

Hohn, E. O., and Sarker, A., K. (1964). Stimulation of mammary lobule alvolar development in the rat by amphenone. *Can. J. Physiol. Pharmacol.* **42,** 683.

Hunt, T. E. (1942). Mitotic activity in the anterior hypophysis of female rats. *Anat. Rec.* **82,** 263.

Hunt, T. E. (1943). Mitotic activity in the anterior hypophysis of female rats of different age groups and at different periods of the day. *Endocrinology* **32,** 334.

Hunt, T. E. (1951). The effect of hypophyseal extract on mitotic activity of the rat hypophysis. *Anat. Rec.* **111,** 713.

Ingle, D. J., and Kendall, E. C. (1937). Atrophy of adrenal cortex of rat produced by administration of large amounts of cortin. *Science* **86,** 245.

Janssens, A. (1965). Effects of injection of cortisol and metyrapone in starved *Xenopus laevis. Gen. Comp. Endocrinol.* **5,** 687.

Jenkins, J. S., Meakin, J. W., Nelson, D. H., and Thorn, G. W. (1958). Inhibition of adrenal steroid 11-oxygenation in the dog. *Science* **128,** 478.

Jenkins, J. S., Meakin, J. W., and Nelson, D. H. (1959). A comparison of the inhibitory effects of 2-methyl-1,2-bis(3-pyridyl)-1-propanone and Amphenone β on the adrenal cortical secretion in the dog. *Endocrinology* **64,** 572.

Kagayama, M. (1965). The follicular cell in the pars distalis of the dog pituitary gland: An electron microscope study. *Endocrinology* **77,** 1053.

Kastin, A. D., Arimura, A., Viosca, S., Barrett, L., and Schally, A. V. (1967). MSH activity in pituitaries of rats exposed to stress. *Neuroendocrinology* **2,** 200.

Katsuki, S., Ito, M., Watanabe, A., Iino, K., Yuji, S., and Kondo, S. (1967). Effect of hypothalamic lesions on pituitary-adrenocortical responses to histamine and methopyrapone. *Endocrinology* **81,** 941.

Kendall, J. W. (1961). Quantitative and temporal studies on the effect of dexamethasone on corticosterone secretion in the rat. *Proc. Soc. Exp. Biol. Med.* **107,** 926.

Kibelstis, J. A., and Ferguson, J. J., Jr. (1964) . Inhibition *in vitro* of adrenal pregnenolone biosynthesis by amphenone and the analogues. *Endocrinology* **74,** 567.

Kief, H., and Schurmever, E., (1956) . Experimentelle untersuchungen über die wirkung synthetischer corticosteriod auf die ratte. *Beitr. Pathol. Anat. Allg. Pathol.* **116,** 541.

Kitay, J. I. (1959). Inhibition of pituitary ACTH release: An extra adrenal action of exogenous ACTH. *Endocrinology* **64,** 475.

Kitay, J. I., Holub, D. A., and Jailer, J. W. (1958). Hormonal regulations of pituitary adrenocorticotropes. *Proc. Soc. Exp. Biol. Med.* **97,** 165.

Kitay, J. I., Holub, D. A., and Jailer, J. W. (1959). Effect of exogenous ACTH on pituitary ACTH concentration after prolonged cortisone treatment and stress. *J. Clin. Invest.* 38, 291.

Knigge, K. M. (1955a). Cytological changes in the hypophysis of intact and thyroidectomized rats subjected to stress. *Anat. Rec.* **122,** 295.

Knigge, K. M. (1955b). Idenity of the argyrophilic cells in the pars distalis of the rat pituitary gland. *Endocrinology* **57,** 719.

Knigge, K. M. (1956). The cellular source of corticotropin. *Anat. Rec.* **124,** 319.

Knigge, K. M. (1957). Cytology and hormone content of rat pituitary glands following adrenalectomy. *Proc. Soc. Exp. Biol. Med.* **94,** 640.

Knutson, F. (1963). Autoradiography of pituitary cells in adrenalectomized rats after injection of adenine-^{14}C *Acta Pathol. Microbiol. Scand.* **58,** 391.

Knutson, F. (1966a). On the cellular site of corticotrophin production. *Acta Endocrinol. (Copenhagen)* **52,** 305.

Knutson, F. (1966b). The rat adenohypophysis during increased corticotrophin production. *Acta Endocrinol. (Copenhagen), Suppl.* **116,** 78 pp.

Kobayashi, Y. (1965). Functional morphology of the pars intermedia of the rat hypophysis as revealed with the electron microscope. *Z. Zellforsch. Mikrosk. Anat.* **68,** 155.

Koneff, A. A. (1944). Effect of ACTH on the anterior pituitary of normal young male rats. *Endocrinology* **34,** 77.

Kracht, J., Breustedt, H. J., and Hachmeister, U. (1965). Immunofluoreszenz der ACTH-bildungsstatten. *Verh. Deut. Ges. Pathol.* **49,** 328.

Kraicer, J., and Cheng, S. (1968). Adenohypophysial DNA and RNA after adrenalectomy, castration and PTU. *Amer. J. Physiol.* **214,** 158.

Kraicer, J., Herlant, M., and Duclos, P. (1967). Changes in adenohypophyseal cytology and nucleic acid content in the rat 32 days after bilateral adrenalectomy and chronic injection of cortisol. *Can. J. Physiol. Pharmacol.* **45,** 947.

Kraulis, I., and Birmingham, M. K. (1963). Hydroxylation sequences in rat adrenal steriods and effects of SU 4885 on C-18 oxygenation. *Fed. Proc., Fed. Amer. Soc. Exp. Biol.* **22,** 530 (abstr.).

Kraus, E. J. (1923). Zur pathologie der basophilen zellen der hypophyse zugleich ein beitrag zur pathologie die morbus basedowi und addisoni. *Virchows Arch. Pathol. Anat. Physiol.* **247,** 421.

Kraus, E. J. (1927). Zur pathologie des morbus addisoni. *Beitr. Pathol. Anat. Allg. Pathol.* **78,** 283.

Krishan, A., Uzman, B. G. and Hedley-Whyte, E. T. (1967). Nuclear bodies: A component of cell nuclei in hamster tissues and human tumors. *J. Ultrastruct. Res.* **19,** 563.

Krulich, L., and McCann, S. M. (1966). Influence of growth hormone on content of growth hormone in the pituitary of normal rats. *Proc. Soc. Exp. Biol. Med.* **121,** 1114.

Kurosumi, K., and Kobayashi, Y. (1966). Corticotrophs in the anterior pituitary glands of normal and adrenalectomized rats as revealed by electron microscopy. *Endocrinology* **78,** 745.

Kurosumi, K., and Oota, Y. (1966). Corticotrophs in the anterior pituitary gland of gonadectomized and thyroidectomized rats as revealed by electron microscopy. *Endocrinology* **79,** 808.

Laqueur, G. L. (1951). Pituitary changes in disease of adrenal cortex. *Amer. J. Pathol.* **27,** 707.

Lehman, J. (1929). Die struktur des hirnanhanges neben nierenloser ratten. *Zr. Gesamte Exp. Med.* **65,** 129.

Leznoff, A., Fishman, J., Talbot, M., McGarry, E. E., Beck, J., and Rose, B. (1962). The cytological localization of ACTH in the human pituitary. *Clin. Invest.* **41,** 1720.

Liddle, G. W., Island, D., Lane, E. M., and Harris, A. P. (1958). Alterations of adrenal steroid patterns in man resulting from treatment with a chemical inhibitor of 11 β-hydroxylation. *J. Clin. Endocrinol. Metab.* **18,** 906.

Lorenzen, L. C., and Ganong, W. F. (1967). Effect of drugs related to α-ethyltryptamine on stress-induced ACTH secretion in the dog. *Endocrinology* **80**, 889.

Lorenzen, L. C., Wise, B. L., and Ganong, W. F. (1965). ACTH inhibiting activity of drugs related to α-ethyltryptamine. *Fed. Proc., Fed. Amer. Soc. Exp. Biol.* **24,** 128. (abstr.).

McCormack, R. V., Reed, C. E., Murray, R. H., and Ray, R. S. (1951). Coexisting acromegaly and Cushing's syndrome. Discussion of hormone production by the pituitary acidophilic cell. *Amer. J. Med.* **10**, 662.

McIntosh, H. W., and Holmes, C. B. (1951). Evidence suggesting suppression of adrenal cortical function by cortisone. *Lancet* **2**, 1061.

McLetchie, N. G. (1942). On staining pituitary basophils of the human hypophysis with reference to Cushing's syndrome. *J. Endocrinol.* **3**, 323.

McNary, W. F. (1957). Progressive cytological changes in the hypophysis associated with endocrine interaction following exposure to cold. *Anat. Rec.* **128**, 233.

Marshall, J. M. (1951). Localization of ACTH by histochemical and immunochemical methods. *J. Exp. Med.* **94**, 21.

Mellgren, J. (1945). The anterior pituitary in hyperfunction of the adrenal cortex. *Acta Pathol. Microbiol. Scand., Suppl.* **60,** 80 pp.

Mellgren, J. (1948). Connection of adreno-genital syndrome with pituitary morphological change. *Acta Path. Microbiol. Scand.* **25**, 284.

Nakane, P. K. (1970). Classifications of anterior pituitary cell types with immunoenzyme histochemistry. *J. Histochem. Cytochem.* **18,** 9.

Nakayama, I., Nickerson, P. A., and Skelton, F. R. (1969). An ultrastructural study of the adrenocorticotropic hormone secreting cell in the rat adenohypophysis during adrenal cortical regeneration. *Lab. Invest.* **21**, 169.

Nichols, J., and Hennigar, G. (1957). Studies on the adrenal during simultaneous administration of DDD and amphenone. *Endocrinology* **61**, 226.

Novikoff, A. B., and Essner, E. (1962). Pathological changes in cytoplasmic organelles. *Fed. Proc., Fed. Amer. Soc. Exp. Biol.* **21**, 1130.

Oka, H. (1937). Die veranderungen der hypophyse bei experimenteller nebenncereninsuffizienz. *Trans. Soc. Pathol. Jap.* **27**, 507.

Palay, S. L. (1958) . The morphology of secretion. In "Frontiers of Cytology" (S. L. Palay, ed.), pp. 335–336. Yale University Press, New Haven, Connecticut.

Pasteels, J. L., and Herlant, M. (1963). Notions nouvelles sur la cytologie de l'antéhypophyse chez le rat. *Z. Zellforsch. Mikrosk. Anat.* **56**, 20.

Pearse, A. G. E. (1953). Cytological and cytochemical investigations of the foetal and adult hypophysis in various physiological and pathological states. *J. Pathol. Bacteriol.* **65**, 355.

Pearse, A. G. E., and van Noorden S. (1963). Functional cytology of the human adenohypophysis. *Can. Med. Ass. J.* **88**, 462.

Perdue, J. F., and McShan, W. H. (1966). Association of adrenocorticotropic hormone activity with small secretory granules from rat anterior pituitary glands. *Endocrinology* **78**, 406.

Purves, H. D. (1956). Hyalinization and basophil adenomata in anterior pituitary. *Proc. Roy. Soc. Med.* **49**, 1014.

Purves, H. D. and Griesbach, W. E. (1956). Changes in basophil cells of rat pituitary after thyroidectomy. *J. Endocrinol.* **13**, 365.

Purves, H. D., and Sirett, N. E. (1965). Assay of corticotrophs in dexamethasone treated rats. *Endocrinology* **77**, 366.

Quenum, A. (1964). Action of Metopirone on the pituitary adenocortical axis of *Cricetomys Gambianus*. *C. R. Soc. Biol.* **158**, 785.

Quenum, A., and Herlant, M. (1964). Etude des cellules corticotropes dans diverse conditions expérimentales chez le rat de gambie. *C. R. Acad. Sci.* **258**, 1902.

Racadot, J. (1963a). Mise en évidence de celles corticotropes d'adénohypophyse chez le chat impubère traité par des inhibiteur de la stéroidogenèse. *C. R. Acad. Sci.* **257**, 1970.

Racadot, J. (1963b). L'origine cytologique de l'hormone corticotrope de l'hypophyse chez le chat impubère. *C. R. Soc. Biol.* **157**, 1931.

Racadot, J., and Herlant, M. (1960). L'action de l'amphenone sur l'hypophyse du chat impubère. *Ann. Endocrinol.* **21**, 828.

Reese, J. D., Koneff, A. A., and Akimoto, M. B. (1939). Anterior pituitary changes following adrenalectomy in the rat. *Anat. Rec.* **75**, 373.

Rennels, E. G. (1964). Electron microscopic alterations in the rat hypophysis after scalding. *Amer. J. Anat.* **114**, 71.

Richard, R. (1965). Estrogen effects on pituitary-adrenal function via the hypothalamus and hypophysis. *Neuroendocrinology* **1**, 322.

Ringler, I., and Brownfield, R. (1960). The thymolytic activities of 16α, 17α-ketals of triamcinolone. *Endocrinology* **66**, 900.

Rochefort, G. J., and Saffran, M. (1957). Distribution of adrenocorticotrophic hormone in the pituitary gland. *Can. J. Biochem. Physiol.* **35**, 471.

Rosenfield, G., and Bascom, W. O. (1956). Inhibition of steroidogenesis by amphenone B. *J. Biol. Chem.* **222**, 565.

Russell, S. M., Dhariwal, A. P., McCann, S. M., and Yates, F. E. (1969). Inhibition by dexamethasone of the *in vivo* pituitary response to corticotropin-releasing factor. *Endocrinology* **85**, 512.

Russfield, A. B. (1955). Endocrine glands after adrenalectomy compared with adrenal insufficiency. *Cancer* **8**, 523.

Russfield, A. B., Reiner, L., and Klaus, H. (1956). The Endocrine significance of hypophyseal tumors in man. *Amer. J. Pathol.* **32**, 1055.

Saffran, M., and Schally, A. V. (1955). *In vitro* bioassay of corticotropin: Modification and statistical treatment. *Endocrinology* **56**, 523.

Salassa, R. M., Kearns, T. P., Kernhan, J. W., Sprague, R. G., and Maccarty, C. S. (1959). Pituitary tumors in patients with Cushing's syndrome. *J. Clin. Endocrinol. Metab.* **19**, 1523.

Sayers, G., and Cheng, C. P. (1949). Adrenalectomy and pituitary adrenocorticotrophic hormone content. *Proc. Soc. Exp. Biol. Med.* **70**, 61.

Sayers, G., and Sayers, M. A. (1947). Regulation of pituitary adrenocorticotrophic activity during the response of the rat to acute stress. *Endocrinology* **40**, 265.

Schelin, U. (1962). Chromophobe and acidophil adenomas of the human pituitary gland. *Acta Pathol. Microbiol. Scand., Suppl.* **158**, 80 pp.

Schelin, U., Lundin, P. M., and Bartholdson, L. (1946). Light and electron microscopic studies on an autonomous stilbestrol-induced pituitary tumor in rats. *Endocrinology* **75**, 893.

Selye, H. (1950). "Stress." Acta, Inc., Montreal.

Severinghaus, A. E. (1937). Cellular changes in the anterior hypophysis with special reference to its secretory activities. *Physiol. Rev.* **17**, 556.

Severinghaus, A. E. (1938). Cytology of the pituitary glands. *Res. Publ., Ass. Res. Nerv. Ment. Dis.* **17**, 69.

Siperstein, E. R. (1963). Identification of the adrenocorticotropin producing cells in the rat hypophysis by autoradiography. *J. Cell Biol.* **17**, 521.

Siperstein, E. R., and Allison, V. P., (1965). Fine structure of the cells responsible for secretion of adrenocorticotrophin in the adrenalectomized rat. *Endocrinology* **76**, 70.

Siperstein, E. R., and Miller, K. J. (1970). Further cytophysiologic evidence for the identity of the cells that produce adrenocorticotrophic hormone. *Endocrinology* **86**, 451.

Slusher, M. A. (1965). Influence of adrenal steroids on self-stimulation in rats. *Proc. Soc. Exp. Med. Biol.* **120**, 617.

Smelser, G. K. (1944). Differential concentration of hormones in the central and peripheral zones of the bovine anterior pituitary gland. *Endocrinology* **34**, 39.

Smith, P. E. (1930). Hypophysectomy and a replacement therapy in the rat. *Amer. J. Anat.* **45**, 205.

Stark, E., Gyevai, A., Acs, Sz. Szalayy, K., Varga, B. (1967). Do corticoids inhibit-ACTH secretion on the hypophyseal level. *Gen. Comp. Endocrinol.* **9**, 494.

Steenburg, R. W. (1965). The influence of depressed circulating cortisol levels on pituitary corticotropin release. *J. Surg. Res.* **5**, 517.

Stigliani, R., Maggi, L., and Fanfani, M. (1954). Richerche sperimentali sulla genesi dell'ormone corticotropo; effetti della somministrazione di granuliipofisari eosinofilie basafili nel ratto ipofisectomizzato. *Arch. Vecchi Anat. Patol.* **22**, 821.

Sussman, L., Librik, L., and Clayton G. (1965). Effect of prior ACTH administration on ACTH release in man. *Metab., Clin. Exp.* **14**, 583.

Swift, H., and Hruban, Z. (1964). Focal degradation as a biological process. *Fed. Proc., Fed. Amer. Soc. Exp. Biol.* **23**, 1026.

Syndor, K. L., and Sayers, G. (1954). Blood and pituitary ACTH in intact and adrenalectomized rats after stress. *Endocrinology* **55**, 621.

Taylor, A. B., Albert, A., and Sprague, R. G. (1949). Adrenotrophic activity of human blood. *Endocrinology* **45**, 335.

Thornton, K. R. (1956). The action of ACTH and cortisone on the anterior pituitary gland. *Proc. Soc. Med.* **49**, 1016.

Tixier-Vidal, A., and Picart, R. (1967). Etude quantitative par radioautographie au miscroscope électronique de l'utilisation de la 4 DL-leucine-^{3}H par les cellules de l'hypophyse du canard en culture organatypique. *J. Cell Biol.* **35**, 501.

Totsuka, S. (1967). Histometrical estimation of the corticotrophin secreting cells in the adrenhypophysis of bilaterally adrenalectomized rats. *Endocrinol. Jap.* **14**, 295.

Tuchmann-Duplessis, H. (1952) . Structural changes in the anterior pituitary with reference to the adrenal cortex. In "Ciba Foundation Colloq. Endocrinol." 4, 33.

Tullner, W. W., and Hertz, R. (1964). Suppression of corticosteroid production in the dog by monase. *Proc. Soc. Exp. Biol. Med.* **116**, 837.

Turner, C. D. and Bagnara, J. T. (1971). In "General Endocrinology" 5th ed. pp. 56–58. Saunders, Philadelphia, Pennsylvania.

Vernikos-Danellis, J. (1965). Effect of stress, adrenalectomy, hypophysectomy and hydrocortisone on the corticotropin-releasing activity of rat median eminence. *Endocrinology* **76**, 122.

Vilar, O., and Tullner, W. (1959). Effects of *o,p*-DDD on histology and 17-hydroxycorticosteroid output of the dog adrenal cortex. *Endocrinology* **65**, 80.

Waxman, S. H., Tippit, D. F., and Kelley, V. C. (1961). The independent estimation of 11-desoxycortisol and cortisol in a single plasma sample. *J. Clin. Endocrinol. Metab.* **21**, 943.

Weiss, M. (1965). The release of pituitary secretion in the platyfish *Xiphophorus maculatus*. *Z. Zellforsch. Mikrosk. Anat.* **68**, 783.

Williamson, D. G., and O'Donnell, V. J. (1967). Mechanism of Metopirone inhibition of a soluble adrenal steroid by 11 β- hydroxylase. *Can. J. Biochem.* **45**, 153.

Yamada, K., and Yamashita, K. (1967). An electron microscopic study on the possible site of production of ACTH in the anterior pituitary of mice. *Z. Zellforsch. Mikrosk. Anat.* **80**, 29.

Chapter 6

Somatotropes

LIGHT MICROSCOPY

Growth hormone is an example of a pituitary hormone that has had its site of production established by association with an observed pathological syndrome. The cell that produces growth hormone was characterized by Benda (1901), who correlated acromegaly with a human eosinophilic adenoma which was characterized by hyperplasia of the acidophils. The administration of extracts from parts of bovine pituitaries rich in acidophils was found to produce accelerated growth in prepubertal rats by Evans and Long (1921) and growth in hypophysectomized tadpoles by Smith and Smith (1923). Smith and MacDowell (1930, 1931) examined the pituitaries of Snell's (1929) dwarf mice and found an absence of acidophils. Ortman (1956) also found a complete lack of somatotropes in dwarf mice, and the electron microscopic findings of Rennels and McNutt (1958) verified the earlier evidence.

As early as 1917 Kojima described the disappearance of somatotropes following conditions of thyroid insufficiency in the rat. In additional studies using rats, Evans *et al.* (1939) noted that following thyroidectomy STH cells disappeared and the animals failed to grow. Ray *et al.* (1950) found that thyroxin administered after hypophysectomy failed to initiate growth. Purves and Griesbach (1952) described two types of acidophils in the rat. The cell they considered to be responsible for the secretion of growth hormone was degranulated during thyroid insufficiency; estrogen treatment or castration had no effect on these somatotropes but did affect the LTH cell. Eartly and LeBlond (1954) reported that if thyroxine secretion was blocked, little or no growth hormone was secreted by the pituitary gland.

Knigge (1958) demonstrated that the disappearance of granules from the α cells preceded disappearance of somatotropic activity. Koneff *et al.* (1948) discovered that administration of somatotropin eventually led to inactivation and involution of STH cells. Administration of potassium iodide to thyroidectomized rats brought about regranulation of previously degranulated somatotropes (Griesbach *et al.*, 1963). Meyer and Evans (1964) administered hydrocortisone to thyroidectomized rats and found that somatotropes became regranulated to 60% of normal after 14 days. These data established the identity of the somatotrope.

Although some controversy has existed concerning staining affinities of somatotropes and mammotropes, there now is general agreement. Dawson (1954, 1963) showed that only somatotropic granules stained orange with orange G stain. With the tetrachrome staining method of Pasteels and Herlant (1963) these acidophils stain orange. Recently Brookes (1967, 1968) used a staining technique that includes azofuchsin, orange G in phosphotungstic acid and wool green S. This stain method stained somatotropic granules orange. Immunofluorescence techniques have been used to distinguish STH cells from other pars distalis cell types in the rat, mouse, and domestic animals by Rümke and Ladiges (1965) and Shiino and Rennels (1966), and in man by Grumbach (1962) and Leznoff *et al.* (1960). Using the Nakane and Pierce (1967) peroxidase-labeled antibody methods, Baker (1970) and Baker *et al.* (1969) were able to differentiate two species of acidophils in the rat.

ULTRASTRUCTURE OF SOMATOTROPES

According to the electron microscopic observations of Rinehart and Farquhar (1953) and Hedinger and Farquhar (1957), somatotropes are the most prevalent and easily identified cell type in the rat pituitary gland. These workers noted secretory granules about 350 mμ in diameter, a Golgi complex, mitochondria with interrupted cristae, and often a well-developed endoplasmic reticulum. This work was confirmed by Lever and Peterson (1960) in the rat; by Barnes (1961, 1962), Yamada and Sano (1960) in the mouse; by Salazar (1963) in the rabbit; by Herlant (1964) in the mole; by Dekker (1968) in the hamster; and by Kurosumi (1961) in various species of fish.

Farquhar and Rinehart (1954a,b) and Yoshimura and Harumiya (1965) confirmed earlier light microscopic findings which indicated that degranulation of somatotropes occurs after thyroidectomy or castration. Rennels and McNutt (1958) and Peterson (1959) both confirmed the absence of somatotropes in dwarf mice pituitaries. In 1967 Yamada *et al.* found hyperplasia and hypertrophy of growth hormone cells in a diabetic strain of mice, thereby adding evidence to the identity of STH cells.

Ultrastructure of Somatotropes in Intact Controls

In our work somatotropes were found to be the most prevalent cell type in the male rat pituitary gland. Fifty to sixty percent of the cells in the male rat and 40% of the cells in the female rat are somatotropes. These cells are found throughout the pituitary gland but especially in the peripheral regions of the gland. Somatotropes are larger in the male rat and the granules also may be larger.

This cell type is easily identified by its medium size, oval shape and abundance of large, round dense secretory granules (Fig. 6.1) throughout the cytoplasm. It usually is found away from capillaries. Sections of secretory granules of the somatotrope vary from 150–400 mμ in diameter with a mean of 235 mμ.

The nuclei of this cell are usually round and rather eccentric in position. The rough endoplasmic reticulum is extensive but usually located at one pole of the cell (Fig. 6.2). The ribosomes are not evenly distributed on the membrane surface and often smooth areas of endoplasmic reticulum are found facing the Golgi apparatus. Whether these areas of smooth endoplasmic reticulum bud off to form small vesicles destined for the Golgi complex, as shown by Jamieson and Palade (1967a,b) in the exocrine pancreas, is a point of conjecture.

One or more Golgi complexes usually are located near the nucleus. Depending on the stage of secretion, it consists of flattened or dilated sacs and tubules (Fig. 6.3). A somatotrope actively engaged in granule formation contains a Golgi system consisting of vesicles as large as 100 mμ in diameter, flattened vacuoles arranged in parallel, and secretory granules in various stages of formation. Multivesicular bodies and lysosomes have been observed in and around the Golgi region which supports the postulation of Novikoff (1957) that lysosomes are formed in the Golgi complex.

Mitochondria in the somatotrope are round or rod-shaped usually with sparse cristae and a rather clear matrix. The cells contain one or perhaps two lysosomes. Centrioles are usually located near the cell surface or near the Golgi region. On occasion it was noted that a cilium (Fig. 6.3) may be associated with centrioles appearing near the cell surface.

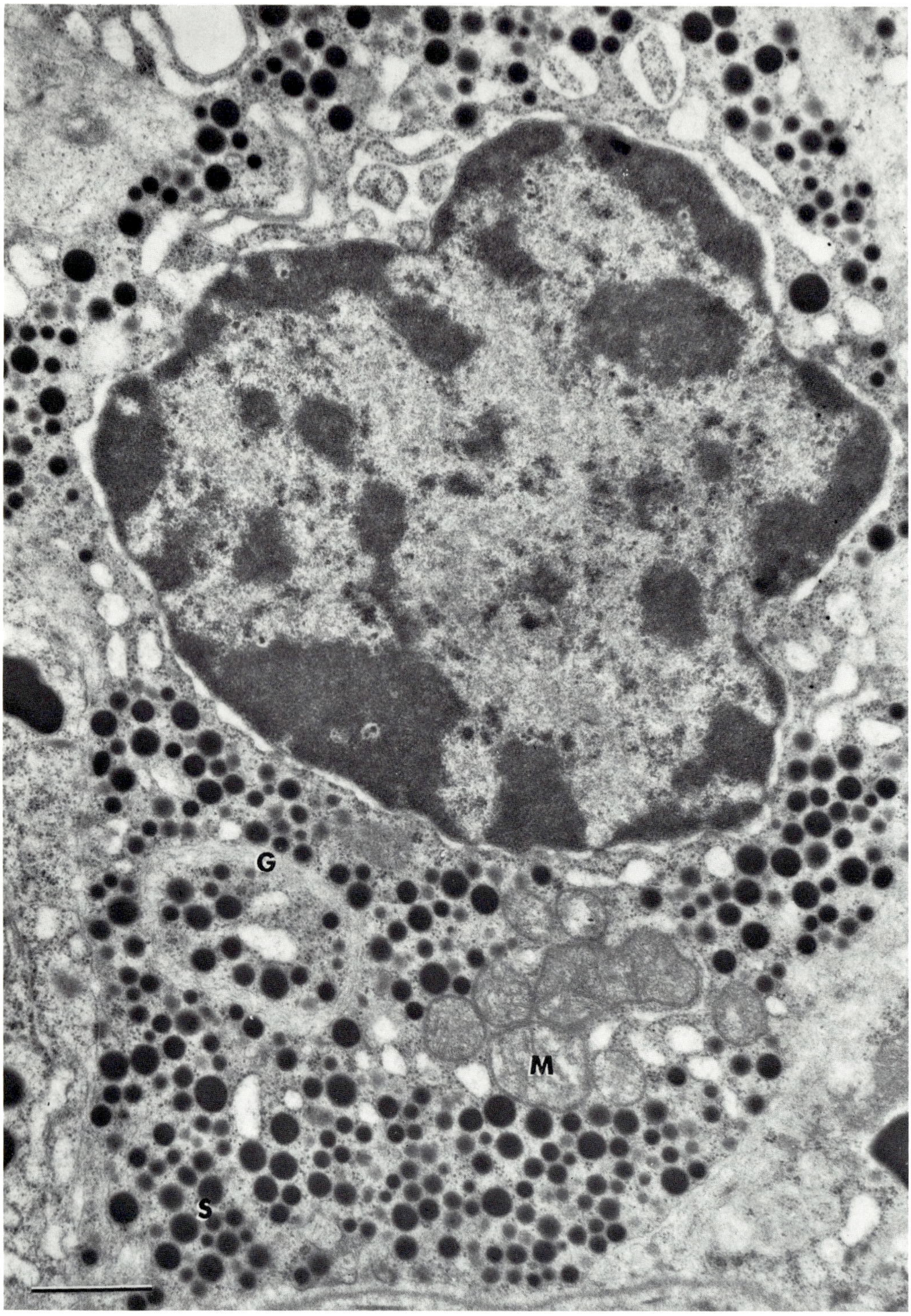

Fig. 6.1. A somatotrope from the pituitary gland of a female rat. A Golgi complex (G), mitochondria (M) with discontinuous cristae and an abundance of dense staining granules (S) are shown. × 18,000. (Glutaraldehyde–paraformaldehyde)

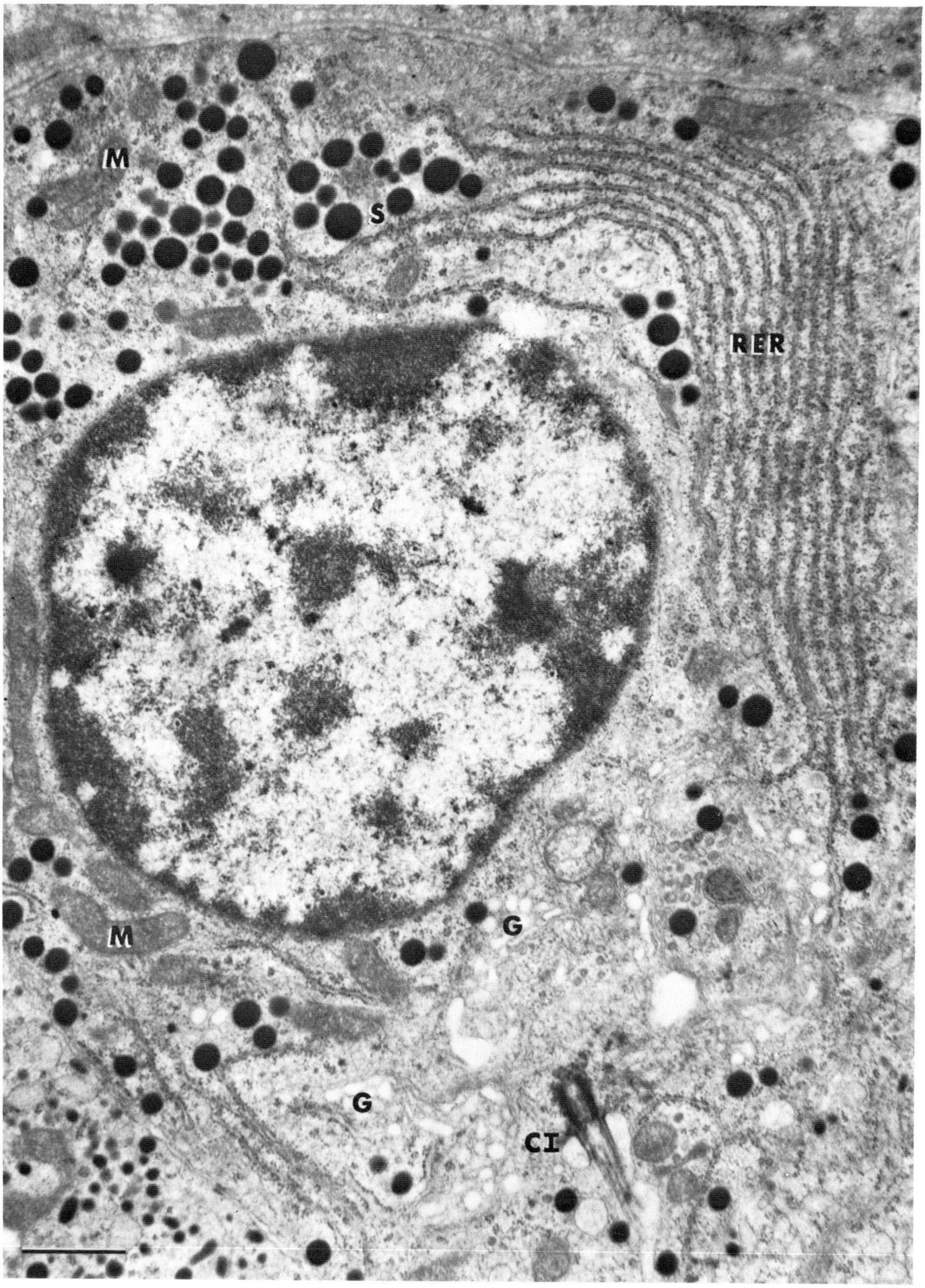

Fig. 6.2. A somatotrope from a male rat pituitary gland. Secretory granules (S) and a lamellar endoplasmic reticulum is shown at one pole of the cell. Dilated Golgi complexes (G), long rodlike mitochondria (M), and a cilium (CI) are also present. × 15,000. (Glutaraldehyde–paraformaldehyde)

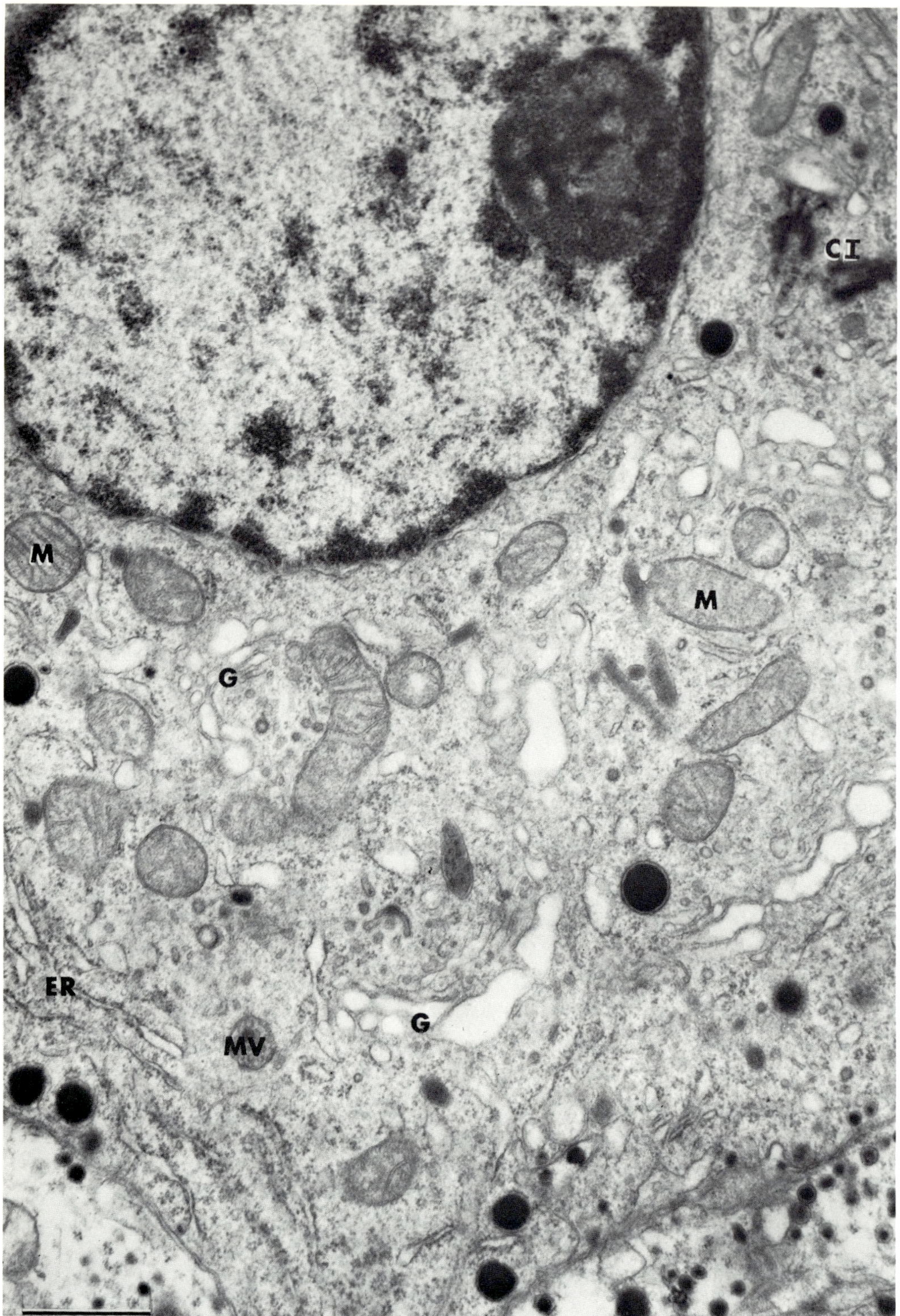

Fig. 6.3. An "Active" somatotrope. The Golgi complexes (G) are greatly dilated. Cilia (CI), multivesicular bodies (MV), mitochondria (M) with discontinuous cristae, and several lamellae of endoplasmic reticulum (ER) are shown. × 20,000. (Glutaraldehyde–paraformaldehyde)

Two types of somatotropes were seen in the adult male pituitary. The active type (Fig. 6.2) had a well-developed rough endoplasmic reticulum, hypertrophied Golgi apparatus, and a few granules at the periphery of the cell. Sometimes granular material could be seen being released at the cell membrane. Active somatotropes are usually found in the central region of the gland. A slightly smaller quiescent, storage-type somatotrope (Fig. 6.1) was more often present in peripheral regions of the adult pituitary gland where the vascular supply is poor. This cell had many granules that obscured observation of the scant Golgi and endoplamic reticulum.

Fine Structure of Somatotropes After Adrenalectomy

Somatotropes appeared rather well granulated up to 5 days following adrenalectomy but later there was an increase in lysosomes. After 14 days as many as 20 lysosomes were observed in some STH cells (Fig. 6.4). These lysosomes showed concentric lamellae, lipid droplets, granules, and other cellular debris within them. Concomitant with the increase in lysosomes was the decrease in granulation in many STH cells. At this time some active somatotropes with well-developed endoplasmic reticulum and Golgi were also present. After 30 days of adrenalectomy about one-third of the somatotropes appeared to be degenerating. These cells had few granules, a faint remnant of endoplasmic reticulum, free ribosomes, an involuted Golgi complex, swollen mitochondria, and a small cytoplasmic region. The somatotropic granules, in general, were smaller in diameter than those in pituitary gland from intact rats. In some cases there appeared to be an increase in cilia in the somatotropes. These degenerating somatotropes would likely be designated as chromophobes by light microscopy.

Ultrastructural Changes in Somatotropes After Other Treatments

Propylthiouracil treatment caused a slight degree of degranulation in somatotropes; and an increase in lysosomes (Fig. 6.5). The somatotropes observed were of the storage type after PTU treatment. Seven days after castration the somatotropes were well granulated but after castration for a longer time some (Fig. 6.6) somatotropes were degranulated. The lysosomes were increased. Somatotropes were degranulated and exhibited swollen mitochondria after adrenalectomy and castration. Metopirone, amphenone, hydrocortisone, and dexamethasone caused degranulation of somatotropes and an increase in lysosomes (Fig. 6.7). The endoplasmic reticulum became more vacuolated and the mitochondria were swollen.

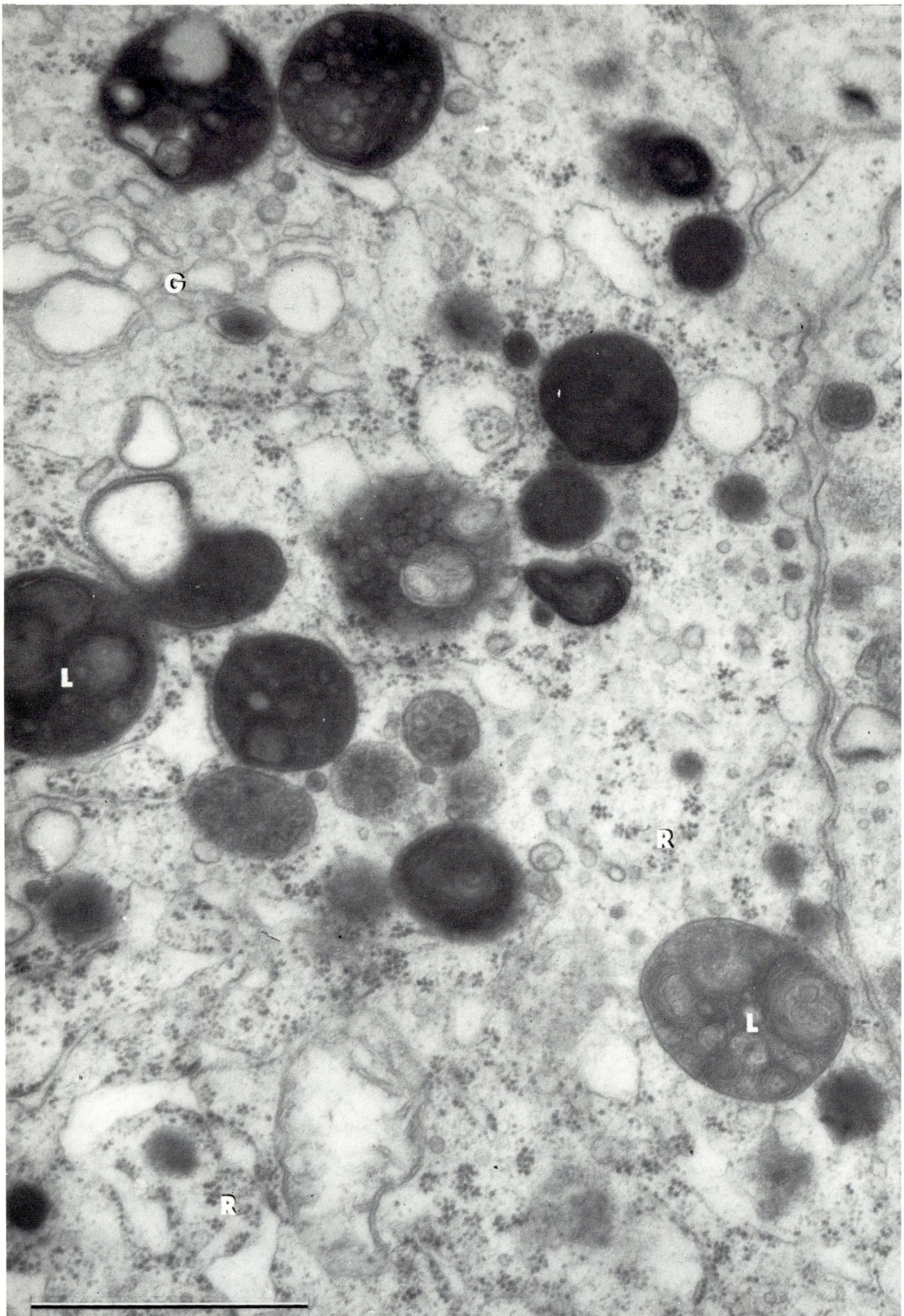

Fig. 6.4. A part of a somatotrope 14 days after adrenalectomy. Many lysosomes (L), a dilated Golgi apparatus (G) (possibly forming lysosomes), and rich ribosomal complexes (R) are present. × 43,500. (Veronal acetate buffered osmic acid)

The somatotropes in the peripheral area of the pituitary showed the most degranulation (see Chap. 5). After ACTH and α-ethyltryptamine administration there were more storage type STH cells.

DISCUSSION OF ULTRASTRUCTURAL FINDINGS

In this study it was found that somatotropes were the most prevalent cell type in the rat pituitary gland and were especially abundant in the peripheral regions. In 1923 Smith and Smith reported high STH activity in the peripheral part of the rat pituitary gland. Rinehart and Farquhar (1953) and Herlant (1964) confirmed by electron and light microscopic studies, respectively, the great abundance of STH cells in the peripheral parts of the pituitary gland. In the present study it was found that there were more somatotropes in the male than the female rat. In this connec-

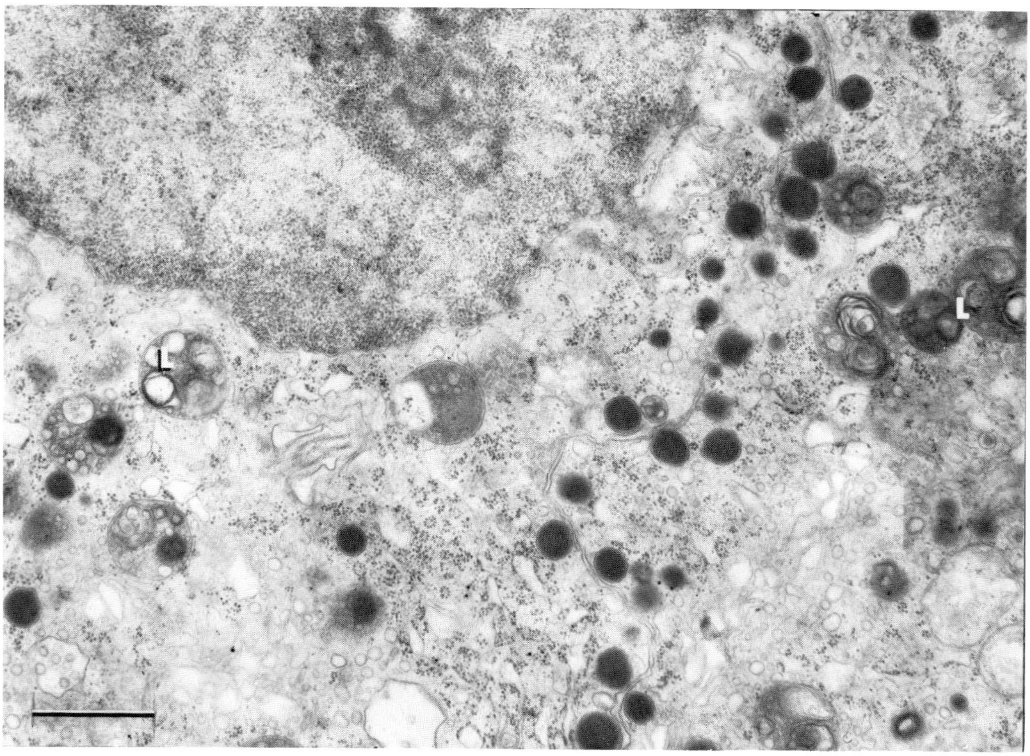

Fig. 6.5. After propylthiouracil treatment some somatotropes showed an increase in lysosomes (L). × 16,500. (Veronal acetate buffered osmic acid)

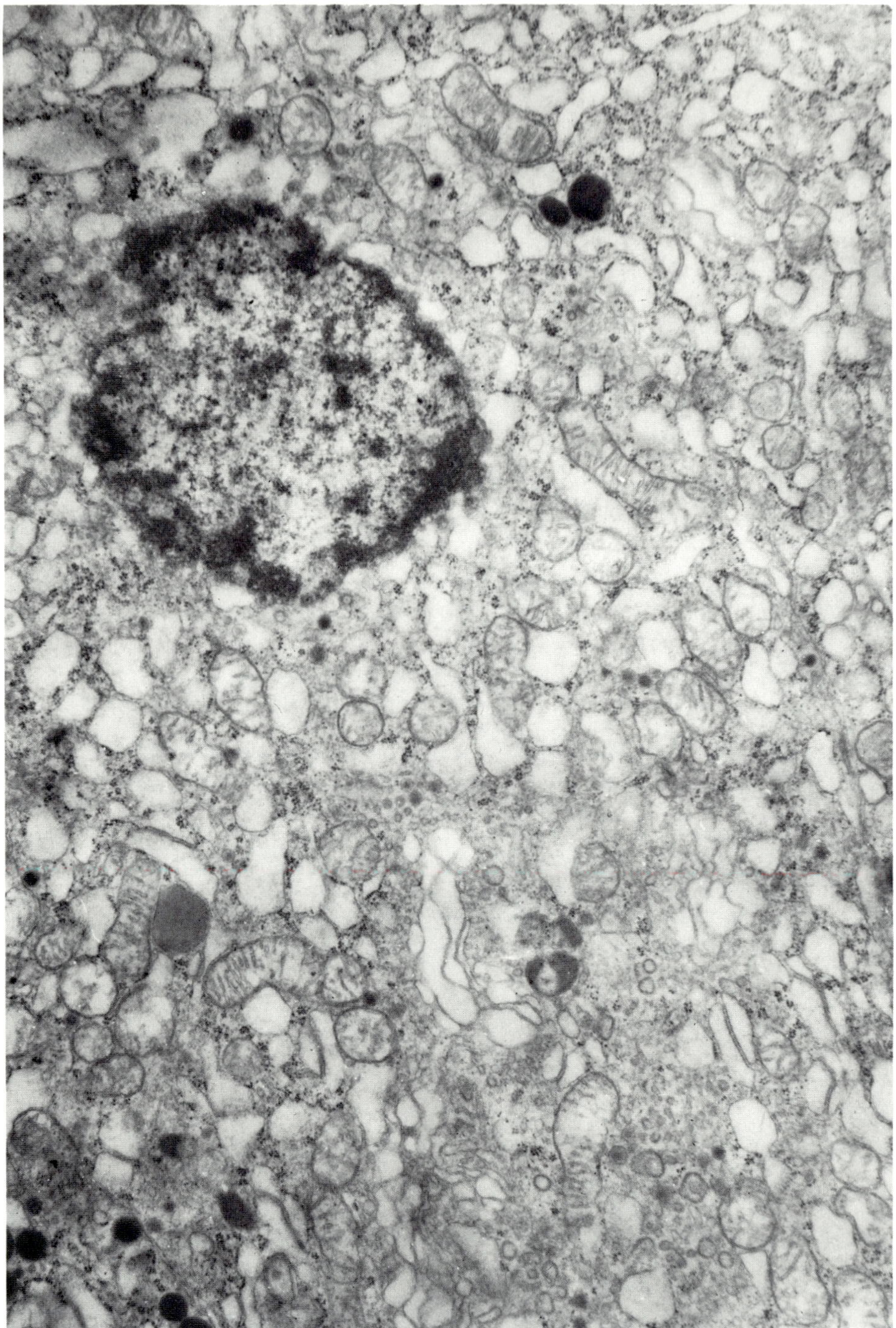

Fig. 6.6. After castration many somatotropes appeared to be degranulated. × 16,500. (Veronal acetate buffered osmic acid)

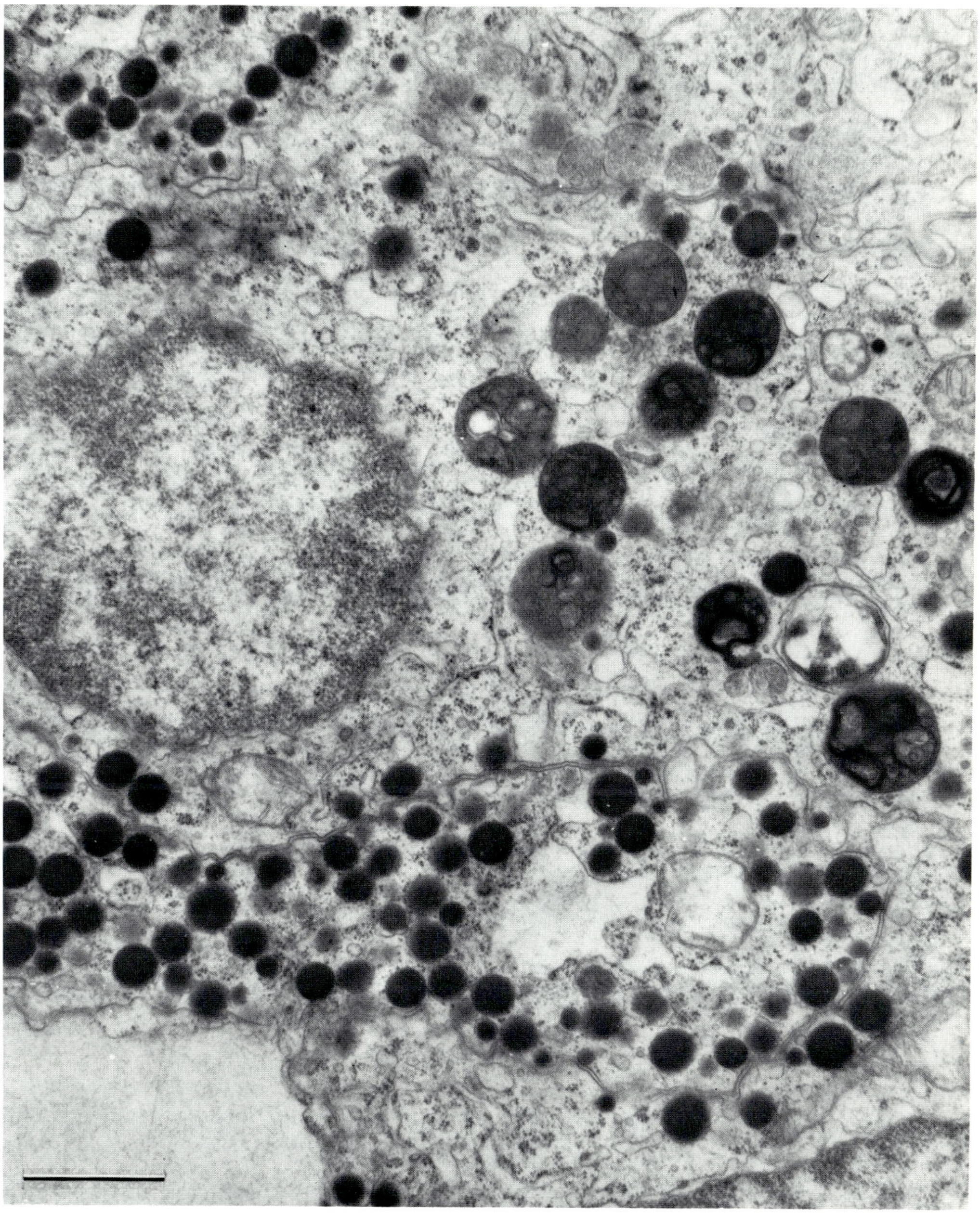

Fig. 6.7. An increase in lysosomes was found in many cells of the pituitary gland, like this somatotrope, after dexamethasone treatment. × 20,000. (Veronal acetate buffered osmic acid)

tion Birge *et al.* (1967) report by radioimmunoassays that the male rat pituitary had more somatotropin than glands from female rats.

Kojima (1917) reported that thyroid hormone was necessary for maintenance of somatotropes because rats treated with prophylthiouracil showed some degranulation of this cell type. Since our treatment with this thyroid blocking agent was of short duration, it can be assumed that a longer treatment would cause a greater degree of degranulation of the STH cells. Purves and Griesbach (1952), Knigge (1958), and Farquhar and Rinehart (1954b) found degranulation of somatotropes following long-term thyroidectomy. Severinghaus (1937) and Farquhar and Rinehart (1954a) observed degranulation of STH cells after castration of rats. In the current studies degranulation of somatotropes was only observed in pituitaries from rats castrated for 2 weeks or longer. Birge *et al.* (1967) also reported a decrease of pituitary STH after castration. In our

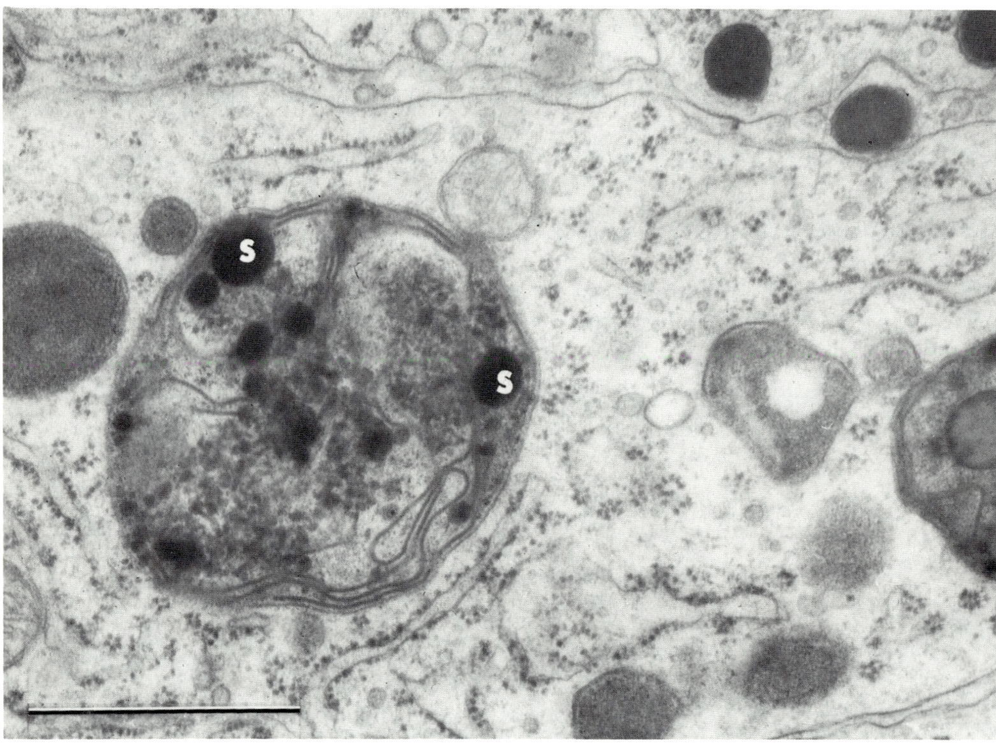

Fig. 6.8. Secretory granules (S) are shown within a lysosome of a somatotrope after adrenalectomy. × 37,500. (Veronal acetate buffered osmic acid)

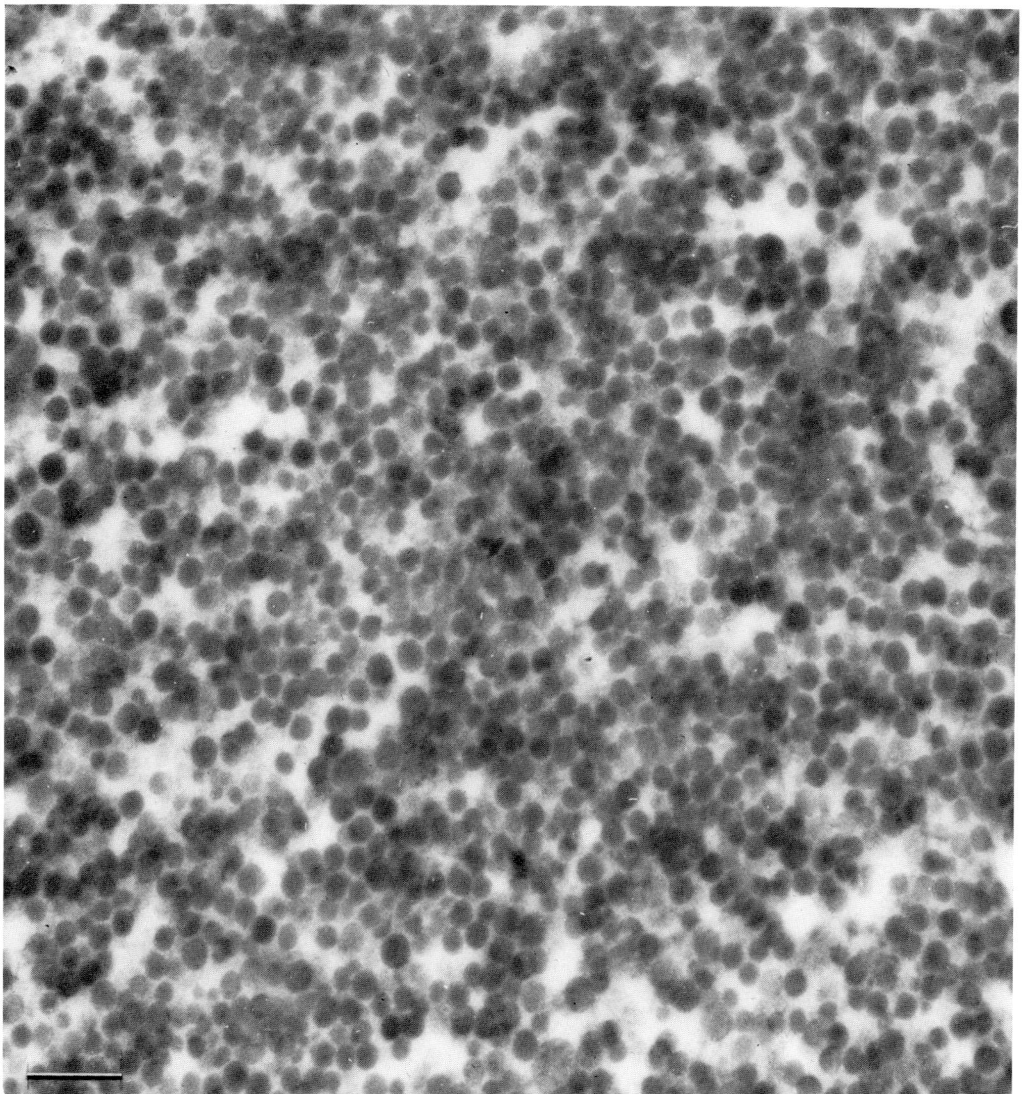

Fig. 6.9. The FE1LSP pellet containing essentially pure granules which were 100–390 mμ in diameter. × 12,000. (Veronal acetate buffered osmic acid)

study the degree of degranulation of STH cells after castration was not as great as those that occurred in somatotropes after long-term adrenalectomy.

The majority of somatotropes from male rat glands were of the storage type, fully granulated and inactive in appearance with poor development of the endoplasmic reticulum and Golgi complex. In 1962 Balogh and Cohen conducted histochemical studies of metabolic and biosynthetic enzyme systems of the human pituitary gland. They showed that acidophils were considerably less active enzymatically than basophils. This suggests that somatotropes are not very active and are of the storage type.

After adrenalectomy for 4 days or longer there were more degranulated somatotropes and they contained a larger number of lysosomes. In pituitaries of adrenalectomized rats lysosomes have phagocytized the secretory granules (Fig. 6.8) of somatotropes. In this way lysosomes may act to regulate the secretory process and prevent overproduction of STH. In a similar situation Smith and Farquhar (1966) showed that after removal of the young from a lactating rat, within 24 hours an abundance of lysosomes were present in the LTH cells; by 72 hours lysosomes and granules were absent from these cells. They concluded that since the mother was no longer suckling, prolactin production or storage was curtailed. The action of lysosomes to phagocytize the prolactin granules is a means of regulating the production of secretory granules.

SOMATOTROPIC ACTIVITIES OF GRANULE FRACTIONS

The somatotropic hormone activities of the isolated granule fractions were determined by use of the tibia epiphyseal plate assay of Greenspan *et al.* (1949). The granule FE1LSP pellet contained 66.6% of the somatotropic activity, whereas 19.3% was found in the soluble SA zone.

The granules in the FE1LSP pellet (Fig. 6.9) had a mean diameter of 242 mμ while those in sections of somatotropes had a mean of 240 mμ. According to the granule measurements most of the STH activity should be in the FE1LSP pellet and the bioassays verified this. These results substantiate the data as reviewed by McShan and Hartley (1965).

Summary

The somatotrope, one of the first pituitary cell types to be identified by histochemical methods, has also been identified by electron microscopy. The degranulation of somatotropes after thyroidectomy, and their absence in dwarf mice have contributed to the identification. The identity of the somatotrope has been further substantiated by our bioassay results and granule measurement studies. Experimentation by Couch *et al.* (1969) have added further evidence to the identification of somatotropes. They injected purified growth hormone releasing factor (GRF) into rats and killed them at different times and fixed the pituitaries for electron microscopy. They noted a great number of granules from somatotropes were being released by exocytosis. The peroxidase-labeled antibody method performed at the electron microscopic level by Nakane (1970) also identified the somatotrope.

REFERENCES

Baker, B. L. (1970). Studies on hormone localization with emphasis on the hypophysis. *J. Histochem. Cytochem.* **18**, 1.

Baker, B. L., Midgley, A. R., Jr., Gersten, B. E., and Yu, Y.-Y. (1969). Differentiation of growth hormone and prolactin containing acidophils with peroxidase-labeled antibody. *Anat. Rec.* **164**, 163.

Balogh, K., and Cohen, R. B. (1962). Histochemical localization of oxidative enzyme systems in the human anterior pituitary. *Endocrinology* **70**, 874.

Barnes, B. C. (1961). Ciliated secretory cells in the pars distalis of the mouse hypophysis. *J. Ultrastruct. Res.* **5**, 453.

Barnes, B. G. (1962). Electron microscopic studies on the cytology of the mouse anterior pituitary. *Endocrinology* **71**, 618.

Benda, C. (1901). Die Microskopischen Befunde bei vier Fallen von Akromegalie. *Deut. Med. Wochenschr.* **27**, 537.

Birge, C. A., Peake, G. T., Mariz, I. K., and Daughaday, W. H. (1967). Radioimmunoassayable growth hormone in the rat pituitary gland: Effect of age, sex, and hormone state. *Endocrinology* **81**, 195.

Brookes, L. D. (1967). A stain for differentiating two types of acidophils in the pituitary. *Gen. Comp. Endocrinol.* **9**, 436.

Brookes, L. D. (1968). A stain for differentiating two types of acidophils in the rat pituitary. *Stan Technol.* **43**, 41.

Couch, E. F., Arimura, A., Schally, A. V., Saitor, M., and Sawano, S. (1969). Electron microscopic studies of somatotrophs of rat pituitary after injection of purified growth hormone releasing factor (GRF). *Endocrinology* **85**, 1084.

Dawson, A. B. (1954). The regional localization of five distinct morphological cell types in the anterior pituitary gland of the rhesus monkey. *Anat. Rec.* **120**, 810.

Dawson, A. B. (1963). The pituitary gland of the short-tailed shrew. *Anat. Rec.* **145**, 315 (abstr.).

Dekker, A. (1968). Electron microscopic study of somatotropic and lactotropic pituitary cells of the Syrian hamster. *Anat. Rec.* **162**, 123.

Eartly, H., and Leblond, C. P. (1954). Identification of the effects of thyroxine mediated by the hypophysis. *Endocrinology* **54**, 249.

Evans, H. M., and Long, J. A. (1921). The effect of the anterior lobe administered intraperitoneally upon growth, maturity and the oestrous cycles of the rat. *Anat. Rec.* **21**, 62.

Evans, H. M., Simpson, M. E., and Pencharz, R. I. (1939). Relation between the growth promoting effects of the pituitary and thyroid hormones. *Endocrinology* **25**, 175.

Farquhar, M. G., and Rinehart, J. F. (1954a). Electron microscopic studies of the anterior pituitary gland of castrate rats. *Endocrinology* **54**, 516.

Farquhar, M. G., and Rinehart, J. F. (1945b) . Cytologic alterations in the anterior pituitary gland following thyroidectomy: An electron microscope study. *Endocrinology* **55**, 857.

Greenspan, F. S., Li, C. H., Simpson, M. E., and Evans, H. M. (1949). Bioassay of hypophyseal growth hormone: The tibia test. *Endocrinology* **45**, 455.

Griesbach, W. E., Evans, E. S., and Chaikoff, I. L. (1963). Does iodide produce significant regranulation of pituitary acidophils in thyroidectomized rats? *Endocrinology* **72**, 474.

Grumbach, M. M. (1962). Intracellular detection of hormones by immunochemical means. *Ciba Found. Colloq. Endocrinol.* **14**, 373.

Hedinger, C. E., and Farquhar, M. G. (1957). Elektronenmikroskopische Untersuchungen von zwei Typen Acidophiler Hypophysenvorderlappenzellen bei der Ratte. *Schweiz. Z. Allg. Pathol. Bakteriol.* **20**, 766.

Herlant, M. (1964). The cells of the adenohypophysis and their functional significance. *Int. Rev. Cytol.* **17**, 299.

Jamieson, J. D., and Palade, G. E. (1967a). Intracellular transport of secretory proteins in the pancreatic exocrine Cell: I. Role of the peripheral elements of the Golgi complex. *J. Cell Biol.* **34**, 577.

Jamieson, J. D., and Palade, G. E. (1967b). Intracellular transport of secretory proteins in the pancreatic exocrine cell. II. Transport of condensing vacuoles and zymogen granules. *J. Cell Biol.* **34**, 579.

Knigge, K. M. (1958). Cytology and growth hormone content of rat pituitary after thyroidectomy and stress. *Anat. Rec.* **130**, 543.

Kojima, M. (1917) . The relationship of the pituitary body with the thyroid and parathyroid and certain other organs of the rat. *Quart. J. Exp. Physiol.* **11**, 319.

Koneff, A. A., Simpson, M. E., Evans, H. M., and Li, C. H. (1948). Gigantism produced in normal rats by injection of pituitary growth hormone: histological changes in pituitary. *Growth* **12**, 33.

Kurosumi, K. (1961). Electron microscopic analysis of the secretion mechanism. *Int. Rev. Cytol.* **11**, 1.

Lever, J. P., and Peterson, R. R. (1960). Cellular identities in the pars distalis of the rat pituitary. *Trans. N. Y. Acad. Sci.* **22**, 504.

Leznoff, A., Fishman, J., Goodfriend, L., McGarry, E., Beck, J., and Rose, B. (1960). Localization of fluorescent antibodies to human growth hormone in human anterior pituitary gland. *Proc. Soc. Exp. Biol. Med.* **104**, 232.

McShan, W. H., and Hartley, M. W. (1965). Production, storage and release of anterior pituitary hormones. *Ergeb. Physiol., Biol. Chem. Exp. Pharmakol.* **56**, 264.

Meyer, Y. N., and Evans, E. S. (1964). Acidophil regranulation and increased growth hormone concentration in the pituitary of thyroidectomized rats after cortisol administration. *Endocrinology* **74**, 784.

Nakane, P. K. (1970). Classifications of anterior pituitary cell types with immunoenzyme histochemistry. *J. Histochem. Cytochem.* **18**, 9.

Nakane, P. K., and Pierce, G. B. (1967). Enzyme-labeled antibodies for light and electron microscopic localization of tissue antigen. *J. Cell Biol.* **33**, 307.

Novikoff, A. B. (1957). Biochemical heterogeneity of the cytoplasmic particles of rat liver. *Symp. Soc. Exp. Biol.* **10**, 92.

Ortman, R. (1956). Study of some cytochemical reactions and of hormone content of adenohypophysis in normal and in genetic dwarf mice. *J. Morphol.* **99**, 417.

Pasteels, J. L., and Herlant, M. (1963). Notions nouvelles sur la cytologie de l'antéhypophyse chez le rat. *Z. Zellforsch. Milrosk. Anat.* **56**, 20.

Peterson, R. R. (1959). Electron microscope observations on the pituitary gland of the dwarf mouse. *Anat. Rec.* **133**, 322 (abstr.).

Purves, H. D., and Griesbach, W. E. (1952). Functional differentiation in the acidophil cells and the gonadotropic basophil cells of the rat pituitary. *Proc. Univ. Otago Med. Sch.* **30**, 27.

Ray, R. D., Asling, C. W., Simpson, M. E., and Evans, H. M. (1950). Effects of thyroxine injections on growth and differentiation of the skeleton of hypophysectomized female rats. *Anat. Rec.* **107**, 253.

Rennels, E. G., and McNutt, W. (1958). Fine structure of anterior pituitary of dwarf mice. *Anat. Rec.* **131**, 591 (abstr.).

Rinehart, J. F., and Farquhar, M. G. (1953). Electron microscopic studies of the anterior pituitary gland. *J. Histochem. Cytochem.* **1**, 93.

Rümke, P., and Ladiges, N. (1965). Comparative immunochemical studies with antisera to sheep prolactin and bovine growth hormone on anterior pituitaries of ox, sheep, rats, and mice. *Z. Zellforsch. Milrosk. Anat.* **67**, 575.

Salazar, H. (1963). The pars distalis of the female rabbit hypophysis: An electron microscopic study. *Anat. Rec.* **147**. 469.

Severinghaus, A. E. (1937). Cellular changes in the anterior hypophysis with special reference to its secretory activities. *Physiol. Rev.* **17**, 556.

Shiino, M., and Rennels, E. G. (1966). Cellular localization of prolactin and growth hormone in the anterior pituitary glands of rat and rabbit. *Tex. Rep. Biol. Med.* **24**, 659.

Smith, P. E., and MacDowell, E. C. (1930). Hypophysectomy and a replacement therapy in the rat. *Amer. J. Anat.* **45**, 205.

Smith, P. E., and MacDowell, E. C. (1931). The differential effect of hereditary mouse dwarfism on the anterior pituitary hormones. *Anat. Rec.* **50**, 85.

Smith, P. E., and Smith, I. P. (1923). The response of the hypophysectomized tadpole to the intraperitoneal injection of the various lobes and colloid of the bovine hypophysis. *Anat. Rec.* **25**, 150.

Smith, R. E., and Farquhar, M. G. (1966). Lysosome function in the regulation of the secretory process in cells of the rat anterior pituitary gland. *J. Cell Biol.* **31**, 319.

Snell, G. D. (1929). Dwarf, a new Mendelian recessive character of the house mouse. *Proc. Nat. Acad. Sci. U. S.* **15**, 733.

Yamada, K., and Sano, N. (1960). Electron microscopic observations of the anterior pituitary of the mouse. *Okajimas Folia Anat. Jap.* **34**, 449.

Yamada, K., Nakamura, M., and Yamashita, K., (1967). Light and electron microscopy of the adenohypophysis of a diabetic strain of mice. *Z. Zellforsch. Mikrosk. Anat.* **79**, 420.

Yoshimura, F., and Harumiya, K. (1965). Electron microscopy of the anterior lobe of pituitary in normal and castrated rats. *Enrocrinol. Jap.* **12**, 119.

Chapter 7

Mammotropes

Lactogenic hormone (LTH) of the anterior pituitary gland was discovered by Stricker and Grueter (1928). Riddle *et al.* (1933) reported that this hormone caused enlargement and milk secretion in pigeon crop glands. Astwood and Fevold (1939) reported the crystalization of LTH. Astwood and Greep (1938) called this hormone a third gonadotropin designated as luteotropin. Evans *et al.* (1941) found that a fraction of lactogenic hormone produced the same affects as luteotropic hormone. It is generally accepted that both lactogenic and luteotropic hormones are one and the same.

LIGHT MICROSCOPY

Erdheim and Stumme (1909) were the first investigators to note a decrease in granulation of certain acidophils during lactation in the rat. Everett and Baker (1945) made differential cell counts from sections of female rat anterior pituitaries during pregancy and lactation. They found that granulation decreases during pregnancy but 3 days postpartum there is an increase in number and granulation of LTH cells. They also noted that during lactation these cells undergo degranulation. Dawson (1946) in the cat and Desclin (1947) in rats found that during lactation these cells undergo degranulation. These investigators also reported degranulation of one population of acidophils and a concomitant increase in riboproteins during pregnancy, pseudopregnancy and lactation in rats.

Purves and Griesbach (1952) observed that LTH acidophils degranulate following castration and estrogen treatment.

Meyer *et al.* (1946) showed that estrogen treatment caused an activation of prolactin cells. Meites and Turner (1948) found a marked increase in LTH content of the pituitary gland after estrogen treatment. It was shown by Lacour (1950) that these cells increase in number following castration of male rats. The lack of differentiation of LTH cells of female mice castrated before sexual maturation was reported by Barnes (1962).

Distinguishing between LTH and STH cells has been difficult in some species. Dawson and Friedgood (1938), using a modified azan stain, described two types of acidophils in cats and rabbits. Later, Dawson (1954) noted that certain acidophils stain carmine with azocarmine stain. Granules of mammotropes stained brick red with tetrachrome staining methods of Pasteels and Herlant (1963). Brookes (1967, 1968) also showed that LTH granules stain red. Barnett *et al.* (1961) found that only granules of LTH cells gave a positive sulfhydryl response. Prolactin cells have been identified by immunofluorescent techniques in the rat, cat, mouse, and sheep by Emmart *et al.* (1963), Rümke and Ladiges (1965), by Emmart *et al.* (1965) in a prolatin tumor, and by Shiino and Rennels in the rabbit (1966). In 1968 Nayak *et al.* found prolactin cells in the pituitaries of cows using immunofluorescence methods. The peroxidase-labeled antibody technique was employed by Baker *et al.* (1969) to distinguish somatotropes and mammotropes.

ULTRASTRUCTURE OF THE MAMMOTROPE IN INTACT AND TREATED RATS

Electron microscopic studies of prolactin cells by Hedinger and Farquhar (1957) revealed elliptically or irregularly shaped granules of 400–800 mμ in diameter. Hymer *et al.* (1961) reported an increase in endoplasmic reticulum and a decrease in granulation of mammotropes after estrogen treatment and during lactation. Bates *et al.* (1962) found that prolactin did not accumulate in a pituitary tumor but was secreted quickly and not stored. Ultrastructural changes occurring in pituitary glands during pregnancy and lactation have been described in mice by Garwood and Latta (1963) and in the rabbit by Young *et al.* (1967). Pasteels (1961, 1962) reported that if a female rat is separated from her young for 10 hours on the sixth day of lactation, the granulation of LTH cells increases greatly, but ½-hour of sucking is sufficient to reduce the pituitary LTH by 96%, presumably because of the massive degranulation that occurs. Tiboldi *et al.* (1967) showed that an estrogen induced pituitary adenomata contained an abundance of prolactin cells that were well granulated. Zambrano and Deis (1970) found that estradiol transplants into the hypothalamus of rats resulted in an activation of mammotropes as exhibited by an increased endoplasmic reticulum and secretory activity.

There are more mammotropes in the male pituitary gland than first reported by early light microscopists. Mammotropes comprise about 10% of the cell types in the male rat pituitary and about 40% in the female. In the male they are the third most abundant cell type, after somatotropes and follicle-stimulating gonadotropes, whereas in the female the luteotropes are almost as abundant as somatotropes. Luteotropes are usually found in the central part of the gland and often these cells are associated with corticotropes.

The LTH cells are often similar in size and shape to somatotropes. However, in the female and sometimes in the male, when the luteotrope is stimulated it may double in size and become more irregular in shape. The feature that made this cell easy to identify was its intensely electron-dense and irregularly shaped granules (Fig. 7.1). Within the Golgi region (Fig. 7.2) the granules may measure only 100 mμ in diameter, but many of the mature granules are 800 mμ in maximal diameter. There are fewer granules in mammotropes than in any other cell type but this varies, depending somewhat on the physiological state of the animal.

Rod-shaped mitrochondria are rather numerous and are usually distributed among the granules. There are more lysosomes in the luteotrope than in the somatotrope. Another feature of this cell that aids in its identification is well-developed and extensive endoplasmic reticulum (Fig. 7.3). When at rest this cell has a closely packed endoplasmic reticulum at one pole but upon stimulation such as during pregnancy (Fig. 7.3) the cisternae become more hypertrophied and extensive. There are more ribosomes attached to the endoplasmic reticulum and fewer free ribosomes than in the somatotrope. The Golgi apparatus is extensive, especially in the pregnant rat (Fig. 7.3). There are usually several Golgi complexes in LTH cells.

Ultrastructural Changes in Mammotropes After Various Treatment

After short-term adrenalectomy there were generally more prolactin cells exhibiting increased granulation. They were more conspicuous and

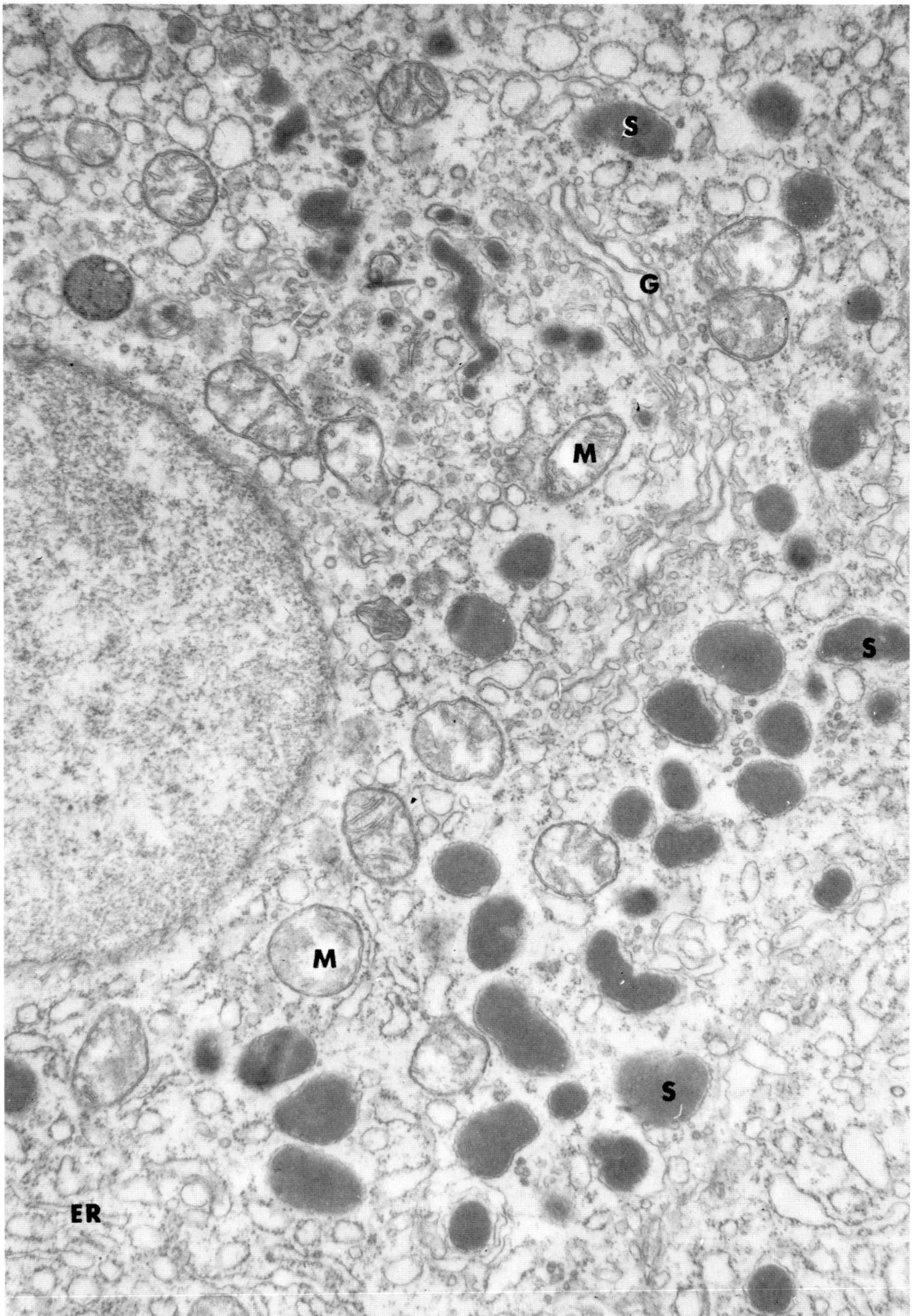

Fig. 7.1. A luteotrope from a male rat anterior pituitary gland. This medium sized cell has granules of different sizes and shapes (S), round or short rod mitochondria (M), an extensive Golgi appartus (G), and an abundance of short cisterane of the endoplasmic reticulum (ER). $\times$ 20,000. (Veronal acetate buffered osmic acid)

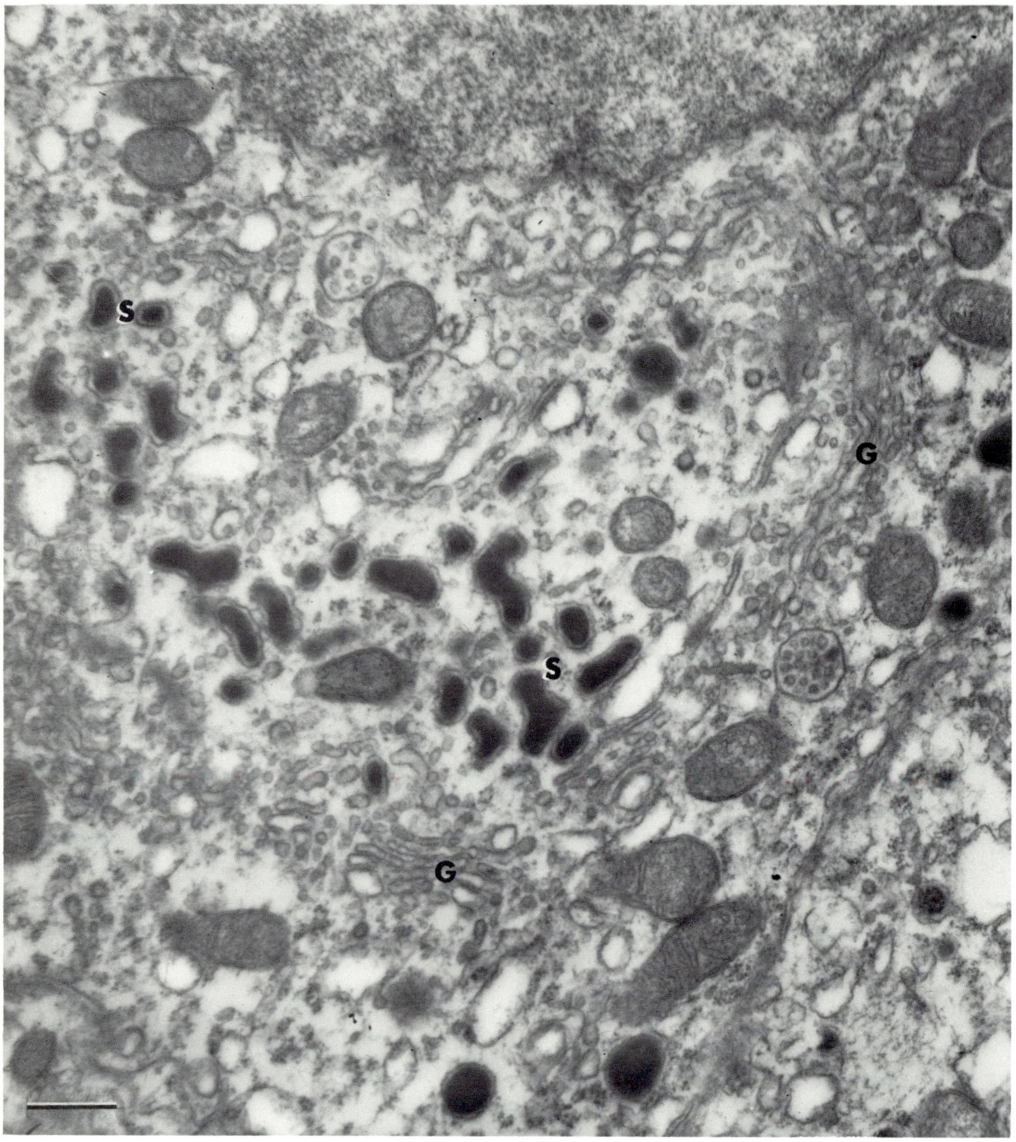

Fig. 7.2. Golgi complexes (G) encircle the nucleus of a luteotrope from a male rat pituitary gland. Granules (S) in different stages of formation are shown. × 12,500. (Veronal acetate buffered osmic acid)

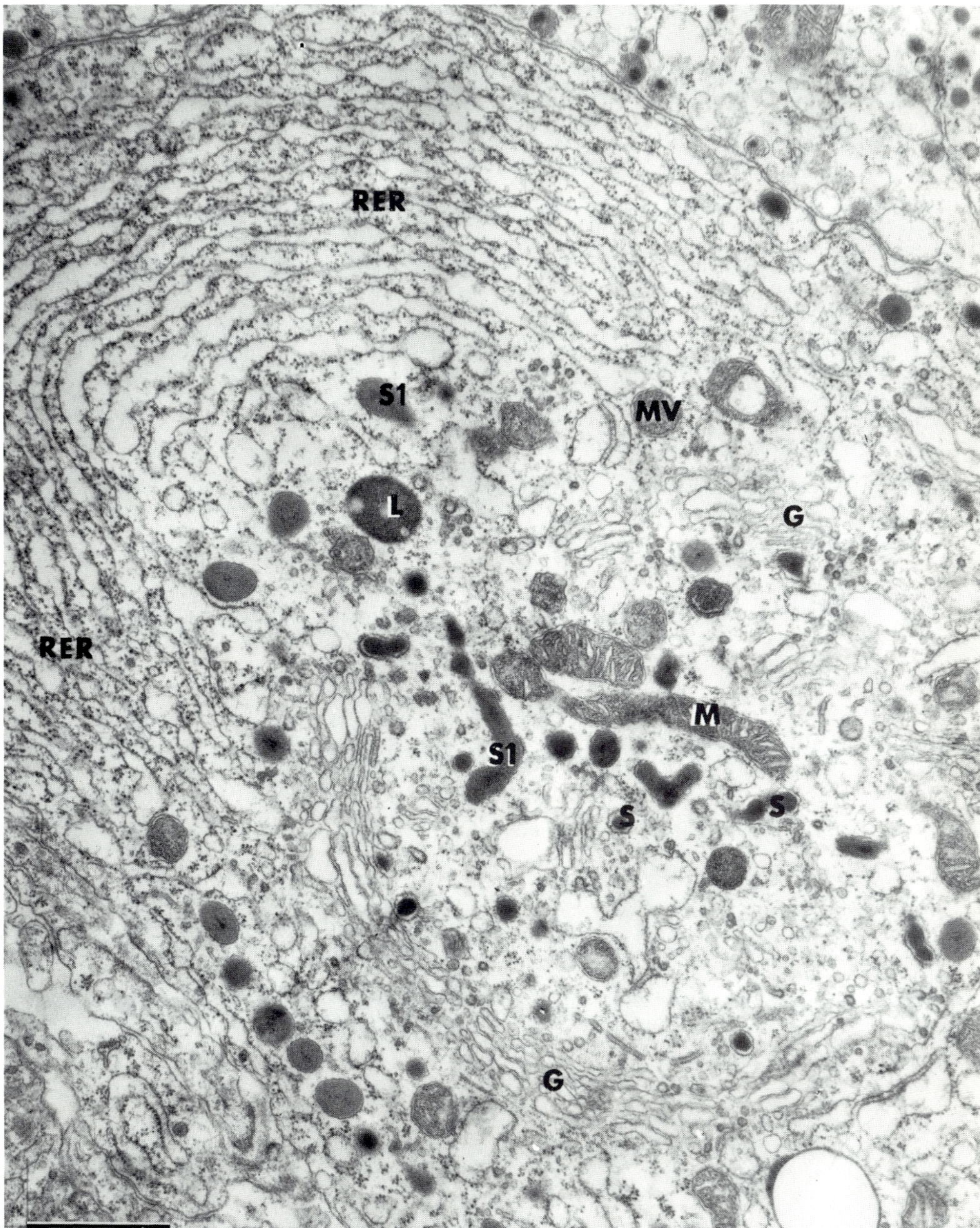

Fig. 7.3. A mammotrope from a pregnant female rat. Extensive rough endoplasmic reticulum (RER) is shown throughout this hypertrophied cell. A lysosome (L), multivesicular body (MV), mitochondria (M) with dense cristae, and several Golgi complexes (G) with granules (S) in various stages of formation are shown. Irregularly shaped mature granules (Sl) are also present. $\times$ 20,000. (Veronal acetate buffered osmic acid.)

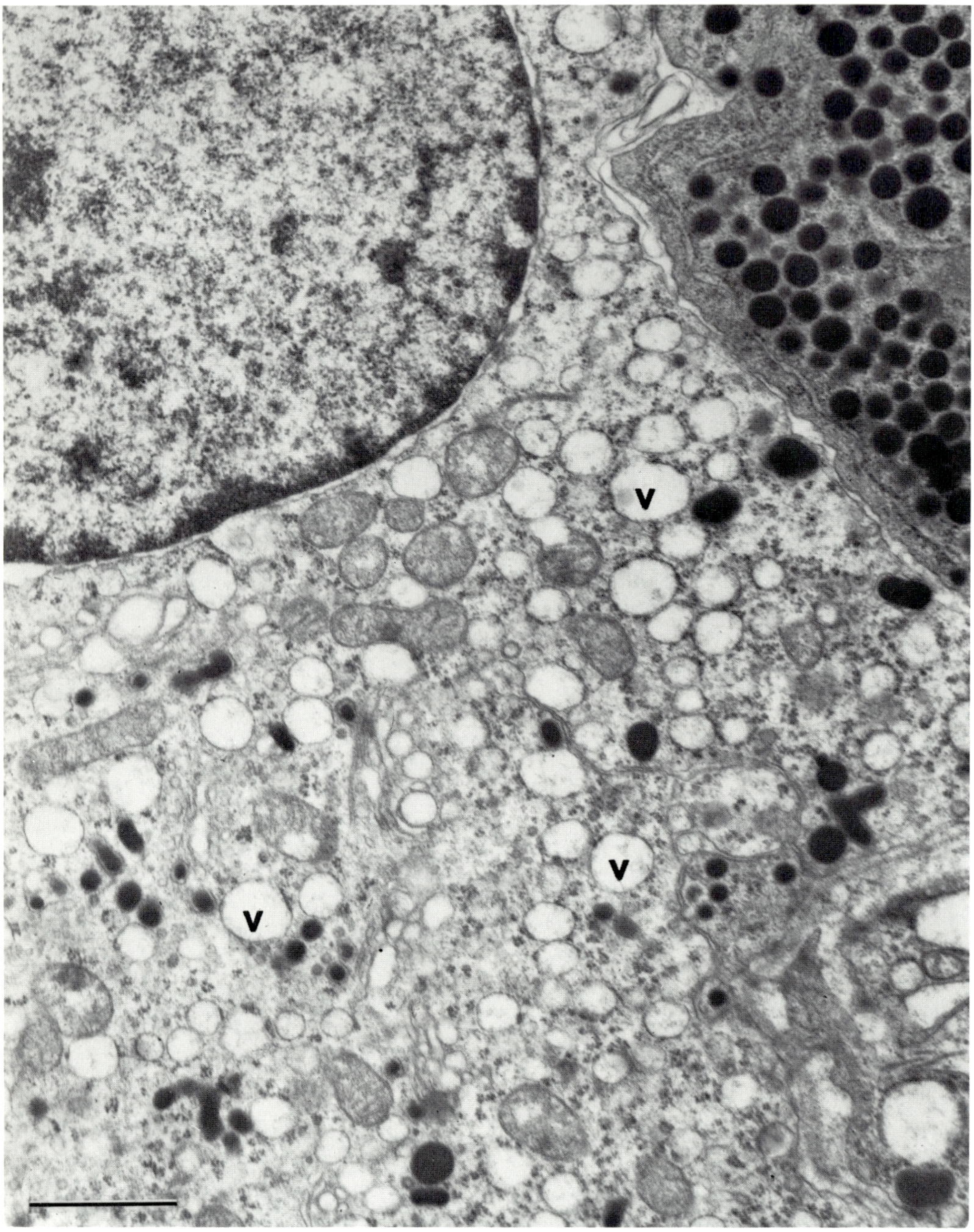

Fig. 7.4. Part of a mammotrope from a female rat during estrus. This cell appears to be degranulated as illustrated by the increase in vacuoles (V). During estrus serum prolactin levels are greatly increased (Amenomori, Y., Chen, C. L., and Meites, J., in *Endocrinology* 86, 506, 1970). × 21,000. (Glutaraldehyde–paraformaldehyde)

contained dilated endoplasmic reticulum after castration or castration and adrenalectomy. The mitochondria of these cells were sometimes swollen and the granulation was usually increased. After the administration of adrenal cortical blockers the number of mammotropes was generally decreased. During the period of estrus some mammotropes appear somewhat degranulated, as is shown in Fig. 7.4.

PITUITARY AUTOGRAFTS

The first evidence of prolactin production by anterior pituitary autografts to the kidney capsule was reported by Desclin (1950). When these rats were injected with estrogen the vaginas became mucified. Vaginal mucification in the presence of estrogen is an effect of progesterone whose secretion is dependent on LTH. Deslin concluded that estrogen stimulated the secretion of LTH.

Everett (1954) found that estrogen stimulation was not necessary for an autograft to produce LTH. Pituitary glands of adult estrus rats were autografted to the kidney capsule and 4 days later the uteri were traumatized. Eight days later decidual tissue was present. Progesterone is required for the formation and maintenance of decidual tissue. Cytological examination of pituitary grafts (Everett, 1956) demonstrated that acidophils were abundant in the grafts. These acidophilic cells were increased in size and number after estrogen treatment. Light and electron microscopy of pituitary grafts under the kidney capsule have been studied by Sanders and Rennels (1959) and Rennels (1962). They found that the only active functional cell contained an abundance of endoplasmic reticulum and large granules; this was assumed to be the LTH cell. Polvliege (1965) and Desclin and Flament-Durant (1966) also reported granulated LTH cells in renal pituitary grafts, but after estrogen treatment these cells were rapidly degranulated. Kovacs (1961) grafted pituitaries to the eye of rats and observed that after 6 weeks only the LTH cells were well granulated.

In tissue cultures of pituitary glands free from the hypothalamus, the LTH responses and granulation are similar to autotransplanted pituitaries (Pasteels and Mulnard, 1961, Robboy and Kahn, 1966). Rao *et al.* (1967) showed incorporation of tritium-labeled amino acids into prolactin cells in bovine anterior pituitary slices grown in culture without the hypothalamus. Tixier-Vidal and Picart (1967) observed that labeled-leucine incorporation occurred only in LTH and MSH cells of cultured pituitary glands. These workers are of the opinion that LTH is stored not only in granules but also in ergastoplasmic cisternae from which it can pass directly into the capillaries and intercellular spaces.

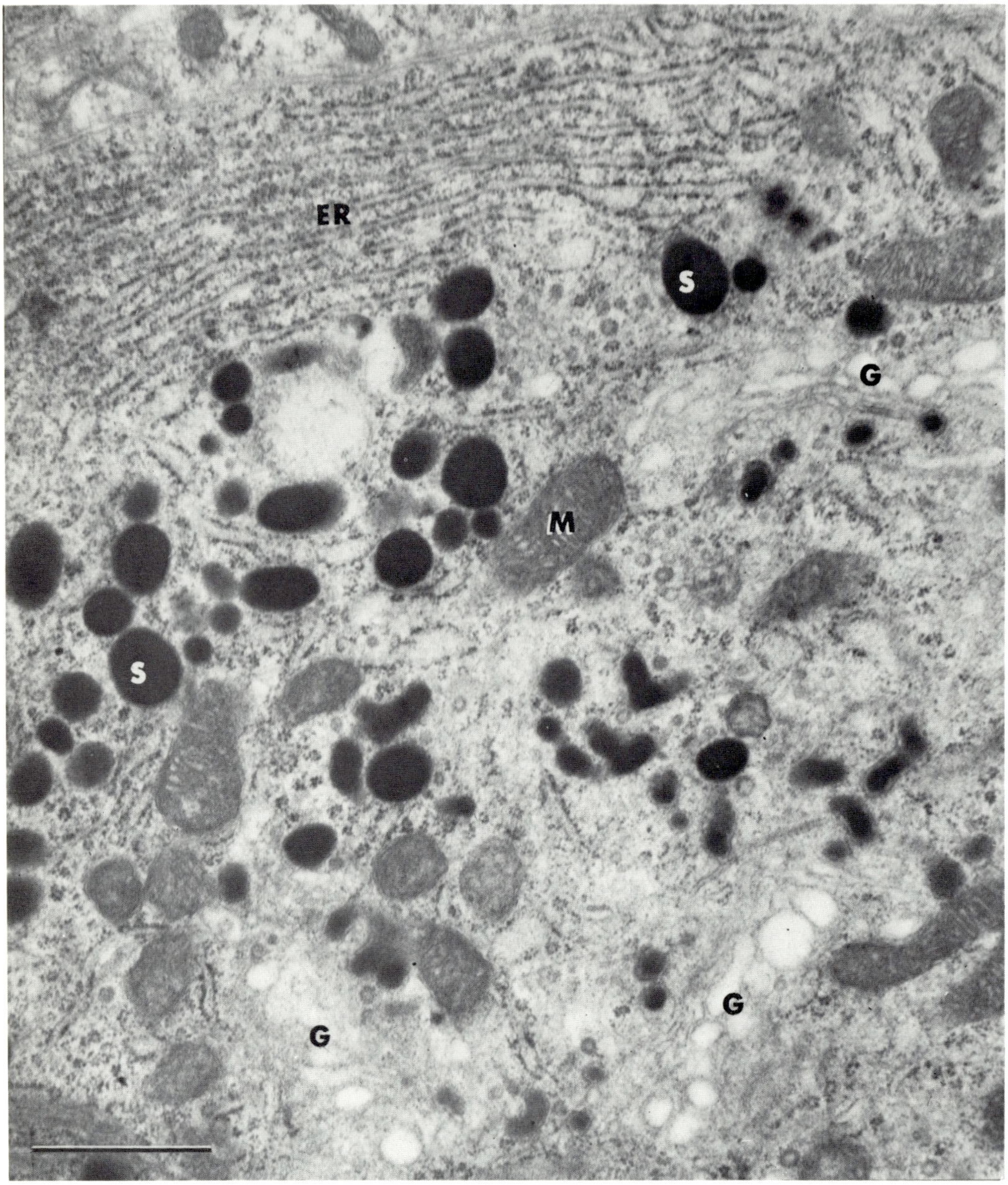

Fig. 7.5. A section of a hypertrophied mammotrope from a female rat with renal pituitary grafts. A well-developed lamellar endoplasmic reticulum (ER), an abundance of secretory granules (S), mitochondria with dense cristae (M), and dilated Golgi complexes (G) are shown. × 25,000. (Glutaraldehyde–paraformaldehyde)

Ultrastructure of The Anterior Pituitary Autografted to the Kidney Capsule in Female Rats

Mammotropes and chromophobes were the primary cell types found in pituitaries autografted under the kidney capsule. The chromophobes appeared more active than in intact controls. They were characterized by having an increased amount of cytoplasm, vesicular endoplasmic reticulum, and increased number of mitochondria with dense matrices. The mammotropes were abundant, comprising perhaps a third to a half of all the cells. These prolactin cells were hypertrophied and showed well-developed lamellar endoplasmic reticulum, hypertrophied Golgi, and many granules (Fig. 7.5) . Other cell types were present but they were smaller (Fig. 7.6) and the Golgi complex and endoplasmic reticulum were poorly developed, indicative of low activity. Although occasionally some active somatotropes (Fig. 7.7) and gonadotropes were present, most of them appear to be inactive. Some cells that resembled melanocyte-stimulating hormone cells (MSH) also appeared well developed. Parts of the graft were necrotic and some cells had an abundance of lysosomes and lipofuchsin droplets.

Discussion

Chen *et al.* (1969) determined the level of LTH secreted by pituitary glands autografted to the kidney to be equal to the greatest amount secreted during the time of highest estrogen secretion in the estrous cycle. Evans and Nikitovitch-Winer (1969) and Evans and Averill (1970) infused median eminence extracts into the renal artery in rats bearing renal autografts. They observed an increase in synthesis and release of gonadotropins and thyrotropin indicating that the involuted cells of the graft could be reactivated.

Mammotropes have been shown to be more prevalent in the female than in the male. These medium-sized cells with irregular shaped granules are located mainly in the medial region of the pituitary gland. They are usually found near corticotropes and appear to be more active after adrenalectomy. The identification of the LTH cells was based on the findings of Nikitovitch-Winer and Everett (1959) and Rennels (1962) who found that only these cells were active and secreting in the pituitaries autografted to the kidney capsule. This conclusion was based on morphological and bioassay data. Also the ultrastructural studies of Hedinger and Farquhar (1957) and Hymer *et al.* (1961) in lactating and estrogen-treated rats have aided in identifying the mammotropic cells.

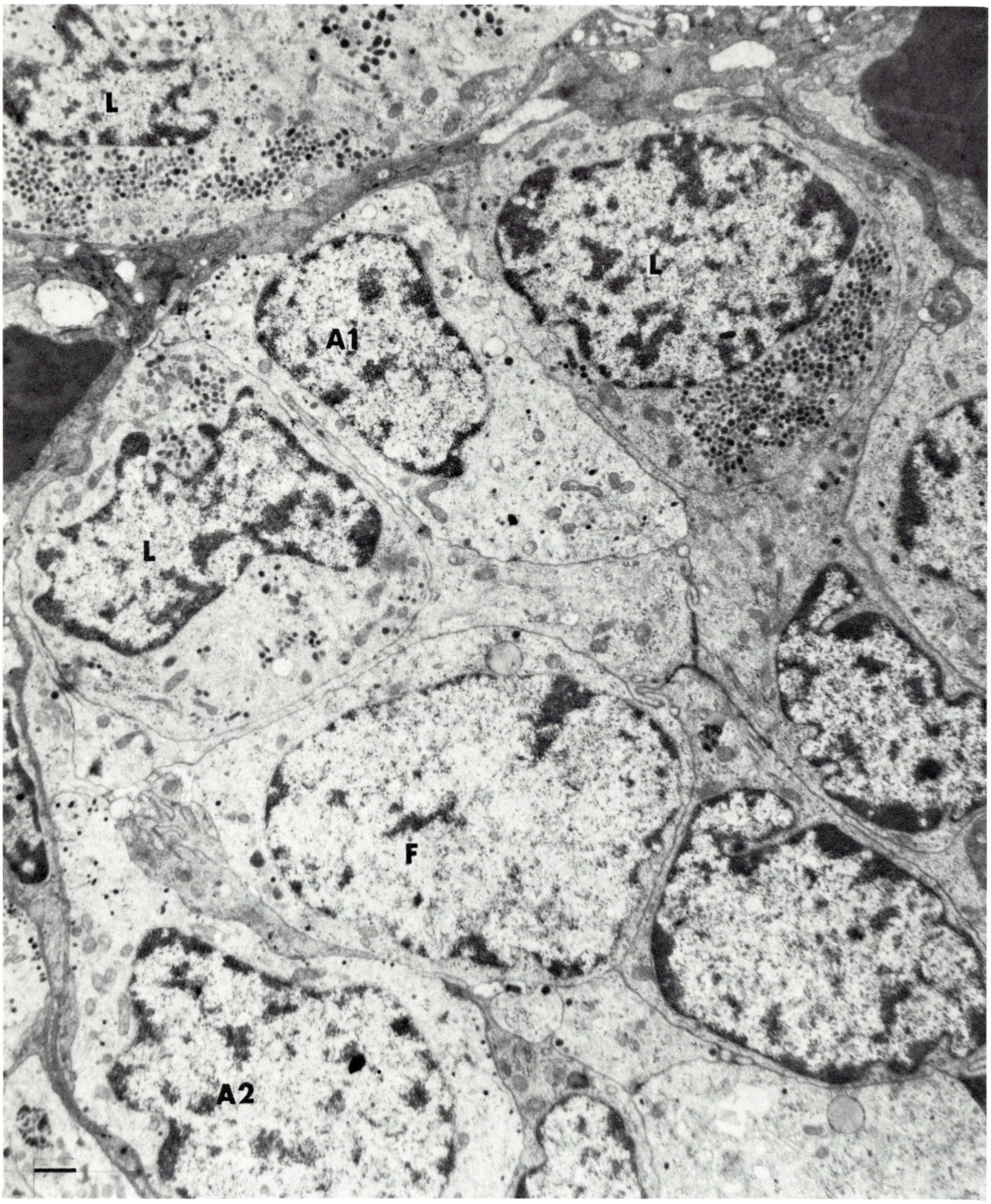

Fig. 7.6. A typical section of pituitary gland autografted to the kidney capsule is shown. It is difficult to make definitive identifications because the cells contain less cytoplasm and appear different from those in intact pituitary glands. Luteinizing hormone gonadotropes (L) appear well granulated. A corticotrope (A1) is shown between the gonadotropes (C) and another one (A2) is next to a follicle cell (F). × 6000. (Glutaraldehyde–paraformaldehyde)

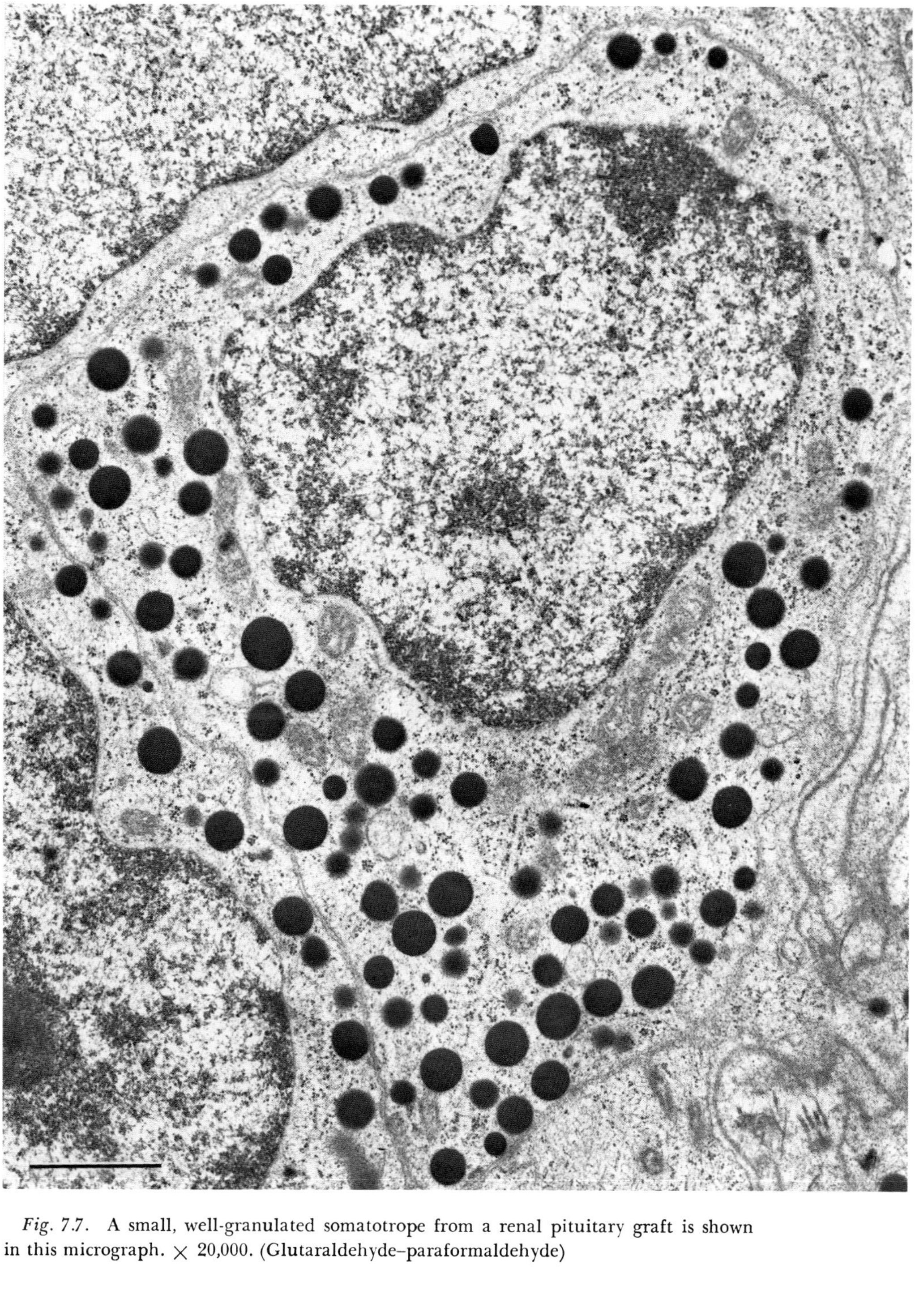

Fig. 7.7. A small, well-granulated somatotrope from a renal pituitary graft is shown in this micrograph. × 20,000. (Glutaraldehyde–paraformaldehyde)

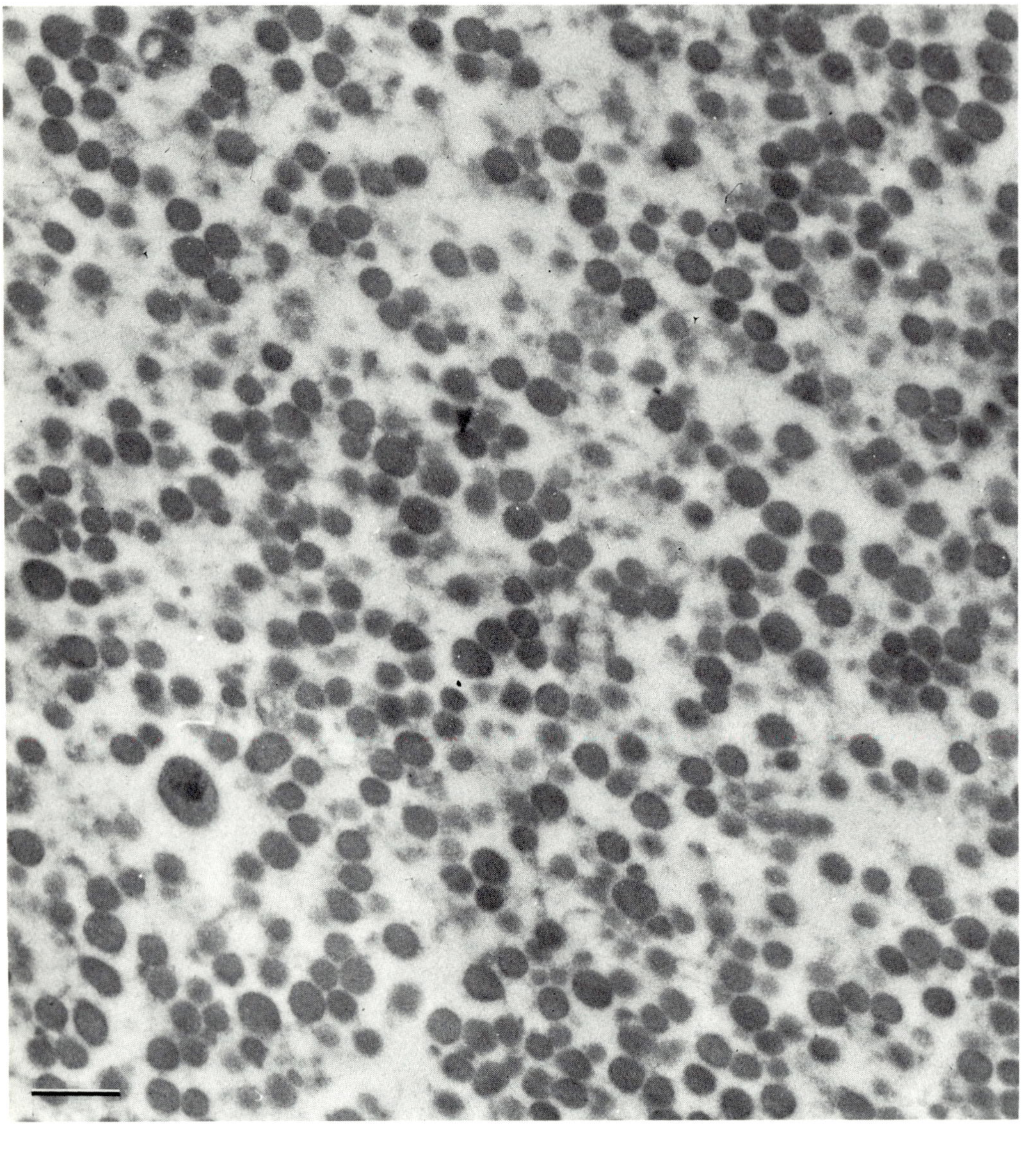

Fig. 7.8. The E2LSP pellet contained stratified granules 220–880 mμ in diameter. $\times$ 12,000. (Veronal acetate buffered osmic acid)

Leutotropic Hormone Activity of Pituitary Granule Fractions

Luteotropin assays were performed using the *in vitro* assay of Mishkinsky *et al.* (1967). This assay, according to the authors, is much more sensitive than the pigeon crop assay.

The E2LSP fraction contained 90.7% of the activity removed from the gradient. There was 8.6% present in the SA soluble hormone zone and only 1% in the FE1LSP, growth hormone zone.

The LTH cells contain granules ranging in diameter from 250–870 mμ with a mean of 318 mμ. Most of the prolactin activity was recovered in the E2LSP zone (Fig. 7.8), which contained granules with a mean diameter of 361 mμ.

Summary

Data from the histochemical, immunofluorescent, and ultrastructural studies of pituitary glands after estrogen administration, during pregnancy, and in renal autografts have contributed to the identification of the mammotropic cell type. The data from the isolation-bioassay and measurement of the granules have provided additional evidence for the establishment of the identity of the mammotrope in the rat pituitary gland.

REFERENCES

Astwood, E. B., and Fevold, H. L. (1939). Action of progesterone on the gonadotropic activity of the pituitary. *Amer. J. Physiol.* **127**, 192.

Astwood, E. B., and Greep, R. O. (1938). A corpus luteum-stimulating substance in the rat placenta. *Proc. Soc. Exp. Biol. Med.* **38**, 713.

Baker, B. L., Midgley, A. R., Jr., Gersten, B. E., and Yu, Y.-Y. (1969). Differentiation of growth hormone and prolactin containing acidophils with peroxidase-labeled antibody. *Anat. Rec.* **164**, 163.

Barnes, B. G. (1962). Electron microscopic studies on the cytology of the mouse anterior pituitary. *Endocrinology* **71**, 618.

Barnett, R. J., Roth, W. D., and Salzer, J. (1961). Histochemical demonstration of sites of LTH in the rat anterior pituitary gland. *Endocrinology* **69**, 1047.

Bates, R. W., Milkovic, S., and Garrison, M. M. (1962). Concentration of prolactin, STH and ACTH in the blood and tumors of rats with transplantable mammotropic pituitary tumors. *Endocrinology* **71**, 943.

Brookes, L. D. (1967). A stain for differentiating two types of acidophils in the pituitary. *Gen. Comp. Endocrinol.* **9**, 436 (abstr.).

Brookes, L. D. (1968). A stain for differentiating two types of acidophils in the rat pituitary gland. *Stain Technol.* **43**, 41.

Chen, C. L., Amenomori, Y., Voogt, J. L., and Meites, J. (1969). Serum prolactin levels in rats bearing pituitary transplants. *Proc. of Endocr. Soc. Meet.* p. 48 (abstr.)

Dawson, A. B. (1946). Some evidences of specific secretory activity of the anterior pituitary gland of the cat. *Amer. J. Anat.* **78**, 347.

Dawson, A. B. (1954). The regional localization of five distinct morphological types of cells in the anterior pituitary of the rhesus monkey. *Anat. Rec.* **120**, 810.

Dawson, A. B., and Friedgood, A. B. (1938). Differentiation of two classes of acidophils in the anterior pituitary of the cat. *Stain Technol.* **13**, 17.

Desclin, L. (1947). Concerning the mechanism of diestrus during lactation in the albino rat. *Endocrinology* **40**, 14.

Desclin, L. (1950). A propos du mécanisme d'action des oestrogènes sur le lobe antérieur du hypophyse chez le rat. *Ann. Endocrinol. (Paris)* **11**, 656.

Desclin, L., and Flament-Durant, J. (1966). Fonction lutéotrophioue de greffe hypophysaires transplantées dans l'hypothalamus chez le rat. *Z. Zellforsch. Mikrosk. Anat.* **69**, 274.

Emmart, E. W., Spicer, S. S., and Bates, R. W. (1963). Localization of prolactin within the pituitary by specific fluorescent antiprolactin globulin. *J. Histochem. Cytochem.* **11**, 365.

Emmart, E. W., Bates, R. W., and Turner, W. A. (1965). Localization of prolactin in rat pituitary and in a transplantable mammotropic pituitary tumor using fluorescent antibody. *J. Histochem. Cytochem.* **13**, 182.

Erdheim, J., and Stumme, E. (1969). Über die Schwangerschaftsveranderung der Hypophyse. *Beitr. Pathol. Anat.* **46**, 1.

Evans, H. M., Simpson, M. E., Lyons, W. R., and Turpeinen, K. (1941). Anterior pituitary hormones which favor the production of traumatic uterine placentomata. *Endocrinology* **28**, 933.

Evans, J. S., and Averill, R. L. (1970). Enhanced thyrotropin secretion by grafted pituitaries infused with hypothalamic extracts. *Fed. Proc., Fed. Amer. Soc. Exp. Biol.* **29**, 311 (abstr.).

Evans, J. S., and Nikitovitch-Winer, M. B. (1969). Functional reactivation and cytological restoration of pituitary grafts by continuous local intravascular infusion of median eminence extracts. *Neuroendocrinology* **4**, 83.

Everett, J. W. (1954). Luteotrophic function of autografts of the rat hypophysis. *Endocrinology* **54**, 685.

Everett, J. W. (1956). Functional corpora lutea maintained for months by autografts of rat hypophysis. *Endocrinology* **58**, 786.

Everett, N. B., and Baker, B. L. (1945). Distribution of cell types in the anterior pituitary during late pregnancy and lactation. *Endocrinology* **37**, 83.

Garwood, V. F., and Latta, J. S. (1963). Electron microscopic studies on the secreting processes in prolatin cells of the mouse anterior pituitary. *Anat. Rec.* **145**, 231.

Hedinger, C. E., and Farquhar, M. G. (1957). Elektronenmikroskopische Untersuchungen von zwei Typen Acidophiler Hypophysenvorderlappenzellen bei der Ratte. *Schweiz. Z. allg. Pathol. Bakteriol.* **20**, 766.

Hymer, W. C., McShan, W. H., and Christiansen, R. G. (1961). Electron microscopic studies of the anterior pituitary gland from lactating and estrogen treated rats. *Endocrinology* **69**, 81.

Kovacs, K. (1961). Histological alterations in the rat pituitary transplanted to the eye. *J. Endocrinol.* **23**, 109.

Lacour, F. (1950). Recherches sur la relation entre les cellules hypophysaires à granulations orangées et les phénomènes de lactation. *C. R. Soc. Biol.* **144**, 248.

Meites, J., and Turner, C. W. (1948). *Mou., Agri. Exp. Sta., Res. Bull.* **415**.

Meyer, R. K., Biddulph, C., and Finerty, J. C. (1946). Pituitary–gonad interaction in immature female parabiotic rats *Endocrinology* **39**, 23.

Mishkinsky, J., Dikstein, S., Ben-David, M., Azeroual, J., and Sulman, F. G. (1967). A sensitive *in vitro* method for prolactin determination. *Proc. Soc. Exp. Biol. Med.* **125**, 360.

Nayak, R., McGarry, E. E., and Beck, J. C. (1968). Site of prolactin in the pituitary gland, as studied by immunofluorescence. *Endocrinology* **83**, 731.

Nikitovitch-Winer, M., and Everett, J. W. (1959). Histocytologic changes in grafts of rat pituitary on the kidney and upon retransplantation under the diencephalon. *Endocrinology* **65**, 357.

Pasteels, J. L. (1961). Premiers resultats de culture combinée *in vitro* d'hypophyse et d'hypothalamus dans le but d'en apprécier la sécrétion de prolactine. *C. R. Acad. Sci.* **253**, 3074.

Pasteels, J. L. (1962). Administration of hypothalamic extracts to the rat pituitary *in vitro* with a view to controlling the secretion of prolactin. *C. R. Acad. Sci.* **254**, 2664.

Pasteels, J. L., and Herlant, M. (1963). New concepts on the cytology of the anterior hypophysis in the rat. *Z. Zellforsch. Mikrosk. Anat.* **56**, 20.

Pasteels, J. L., and Mulnard, J. (1961) . Caractéres morphologiques d'explants hypophysaires cultivés *in vitro*. *C. R. Acad. Sci.* **253**, 1985.

Polvliege, P. (1965). L'influence d'oestrogènes sur l'ultrastructure de greffons hypophysaires chez la ratte. *J. Microsc. (Paris)* **4**, 485.

Purves, H. D., and Griesbach, W. E. (1952). Functional differentiation in the acidophil cells and the gonadotropic basophil cells of the rat pituitary. *Proc. Univ. Otago Med. Sch.* **30**, 27.

Rao, P. M., Robertson, M. C., Winnick, M., and Winnick T. (1967). Biosynthesis of prolactin and growth hormone in slices of bovine anterior pituitary tissue. *Endocrinology* **80**, 1111.

Rennels, E. G. (1962). An electron microscope study of pituitary autograft cells in the rat. *Endocrinology* **71**, 713.

Riddle, O., Bates, R. W., and Dykshorn, S. W. (1933). The preparation, identification and assay of prolactin—A hormone of the anterior pituitary gland. *Amer. J. Physiol.* **105**, 191.

Robboy, S. J., and Kahn, R. (1966). Zone electrophoresis and prolactin activity of rat adenohypophysis cultivated *in vitro*. *Endocrinology* **78**, 440.

Rümke, P., and Ladiges, N. (1965). Comparative immunochemical studies with antisera to sheep prolactin and bovine growth hormone on anterior pituitaries of ox, sheep, rats and mice. *Z. Zellforsch. Mikrosk. Anat.* **67**, 575.

Sanders, A. E., and Rennels, E. G. (1959). Evidence on the cellular source of luteotropin derived from a study of rat pituitary autografts. *Z. Zellforsch. Mikrosk. Anat.* **49**, 263.

Shiino, M., and Rennels, E. G. (1966). Cellular localizations of prolactin and growth hormone in the anterior pituitary glands of the rat and rabbit. *Tex. Rep. Biol. Med.* **24**, 659.

Stricker, P., and Grueter, F. (1928). Action du lobe antérieur de l'hypophyse sur la montée laiteuse. *C. R. Soc. Biol.* **99**, 1978.

Tiboldi, T., Viragh, S., Kovac, K., Hodi, M., and Julesz, M. (1967). Electronmicroscopic studies of oestrogen induced adenomata developing in intersellar and transplanted rat anterior pituitary. *J. Mikros.* **6**, 677.

Tixier-Vidal, A., and Picart, R. (1967). Etude quantitative par radioautographie an microscope électronique de l'activation de la DL-leucine ^{3}H par les cellules de l'hypophyse du canard en culture organatypique. *J. Cell Biol.* **35**, 501.

Young, B. A., Foster, C. L., and Cameron, E. (1967). Ultrastructural changes in the adenohypophysis of pregnant and lactating rabbits. *J. Endocrinol.* **39**, 437.

Zambrano, D., and Deis, R. P. (1970). The adenohypophysis of female rats after hypothalamic oestradiol implants: An electron microscopic study. *J. Endocrinol.* **47**, **101**.

Chapter 8

Chromophobes

Chromophobes are cells that lack staining qualities. This is in contrast to the previously mentioned chromophils which contain granules that stain and amphophils that stain with acidic and basic dyes. No specific hormone has been associated with this cell type since the so-called ACTH chromophobe has been shown to contain fine acidophilic granules.

Rasmussen (1929a,b, 1933) examined many human pituitaries and noted that 52% of the cells were chromophobes, 37% acidophils, and 11% basophils. This high proportion of chromophobes has not been substantiated by electron microscopy. Rinehart and Farquhar (1953) suggested that many cells identified as chromophobes by light microscopy were poorly granulated cells that did not stain intensively. The proportion of chromophobes is in reality far less than previously indicated. Basophilic and fine ACTH acidophilic granules are not very electron dense and the percentage of these cell types actually are greater than indicated by light microscopy.

Since mitotic figures are rarely seen in the pars distalis (Hunt, 1942, 1951), questions arise concerning how cell proportions change and how new cells are derived. Two possibilities were reported by Severinghaus (1933) : *(1)* there are three distinct cell types: acidophils, basophils, and chromophobes, and (2) these cell types represent various forms or physiological stages of the same cell. According to results of electron microscopic studies, the latter does not appear to be the case.

Severinghaus (1933, 1935) claimed to have established a definite relationship between chromophobes and chromophils. He distinguished acidophils from basophils by observing differences in their Golgi complexes. Upon close microscopic examination of the Golgi apparatus in chromophobes it was possible to classify them as either acidophilic or basophilic progeny. He also distinguished between the basophilic and acidophilic chromophobes in the human by differential staining of the nuclei.

In 1939 Severinghaus established a relationship between chromophobes and chromophils. He reported that in the embryonic gland the acidophilic and basophilic chromophobes differentiate from embryonic parenchyma stem cells, and then the acidophilic and basophilic chromophobes differentiate into their respective chromophilic types. It was further proposed that when these acidophils and basophils degranulate they return to their chromophobe state and then differentiate into different kinds of chromophilic cells.

The presence of amphophils (cells having acidic and basophilic staining qualities) might appear to contradict the postulations of Severinghaus (1933). Severinghaus discussed this situation in his review of 1937. Some workers incorrectly classified highly active basophils as acidophils because during heightened activity there was a depletion of basophilic granules. Mitochondria of these basophilic cells often stained with acid dyes and were mistaken for acidophilic mitochondria and are considered to be an amphophilic cell. Amphophils of this kind were more often found in pathological pituitary glands. Although there appears to be some contradictory evidence, it seems likely that chromophobes may be stem cells that differentiate into acidophils or basophils depending on the physiological conditions.

ELECTRON MICROSCOPY OF CHROMOPHOBES

In 1953 Rinehart and Farquhar made an electron microscopic study of the rat anterior pituitary gland. They were the first to describe, on the ultrastructural level, cells devoid of granules, which were designated as chromophobes; this work substantiated the earlier study by Fernández-Morán and Luft (1949). They described these cells as having a vesicular nucleus, rather clear cytoplasm, and a few mitochondria. Electron microscopic studies of mouse pituitary glands by Yamada *et al.* (1960) showed that chromophobes increased in number with advancing age. Schelin (1962) in man, and Holmes and Mandl (1961) in the rat have conducted light and electron microscopic studies of pituitary chromophobe adenomas. Even though these tumors are called chromophobe adenomata, the prominent cell is the one that contains fine granules. This cell may be a corticotrope and not a chromophobe. Chromophobes can be divided into small, medium, stellate, and follicle types.

Small Chromophobes

These cells are characterized by very little cytoplasm surrounding a prominent nucleus. They are slightly smaller than thyrotropes. This type of chromophobe (Fig. 8.1) is devoid of granules but may have a few scattered rough endoplasmic reticular vesicles and some free ribosomes. A few mitochondria with incomplete cristae are present but a Golgi area is absent.

Perhaps these small size chromophobes are stem cells that are not active. Siperstein (1963) reported that after adrenalectomy the number of small chromophobes decreased and that of larger chromophobes increased. Because the larger chromophobes incorporated more labeled amino acid than the other cell types, this cell was designated as the adrenalectomy cell. This study suggests that smaller chromophobes develop into larger chromophobes. Since these large chromophobes are probably chromophils with fine granules, it can be assumed that the small chromophobe differentiates into a chromophilic cell, the ACTH cell. Severinghaus (1937), Ezrin *et al.* (1959), Rovit and Berry (1965), and Capen and Koestner (1967) have also postulated that chromophobes may give rise to chromphils.

Further proof that the small chromophobes are stem cells is provided by the work of Yoshimura *et al.* (1969). Using a gradient centrifugation technique they obtained separate populations of acidophils, basophils, and chromophobes. The pellet containing chromophobes was homotransplanted to the medial basal hypothalamus of hypophysectomized rats. Twenty-four hours to 7 days after transplantation the chromophobes proliferated to some degree by mitosis. Animals were sacrificed at different intervals from 36 hours to 30 days after transplantation, and it was found that chromophobes gradually differentiated into acidophils and basophils. This work indicates that chromophobes can differentiate into chromophils. Also Schechter (1971) has shown in an ultrastructural study of fetal rabbit pituitary glands that chromophobes differentiate into chromophils.

Medium Chromophobe

This chromophobe is medium in size, and contains 2 or 3 times more cytoplasm than small chromophobes (Fig. 8.2). This cell type usually contains a dense cytoplasmic matrix, one Golgi complex that usually ap-

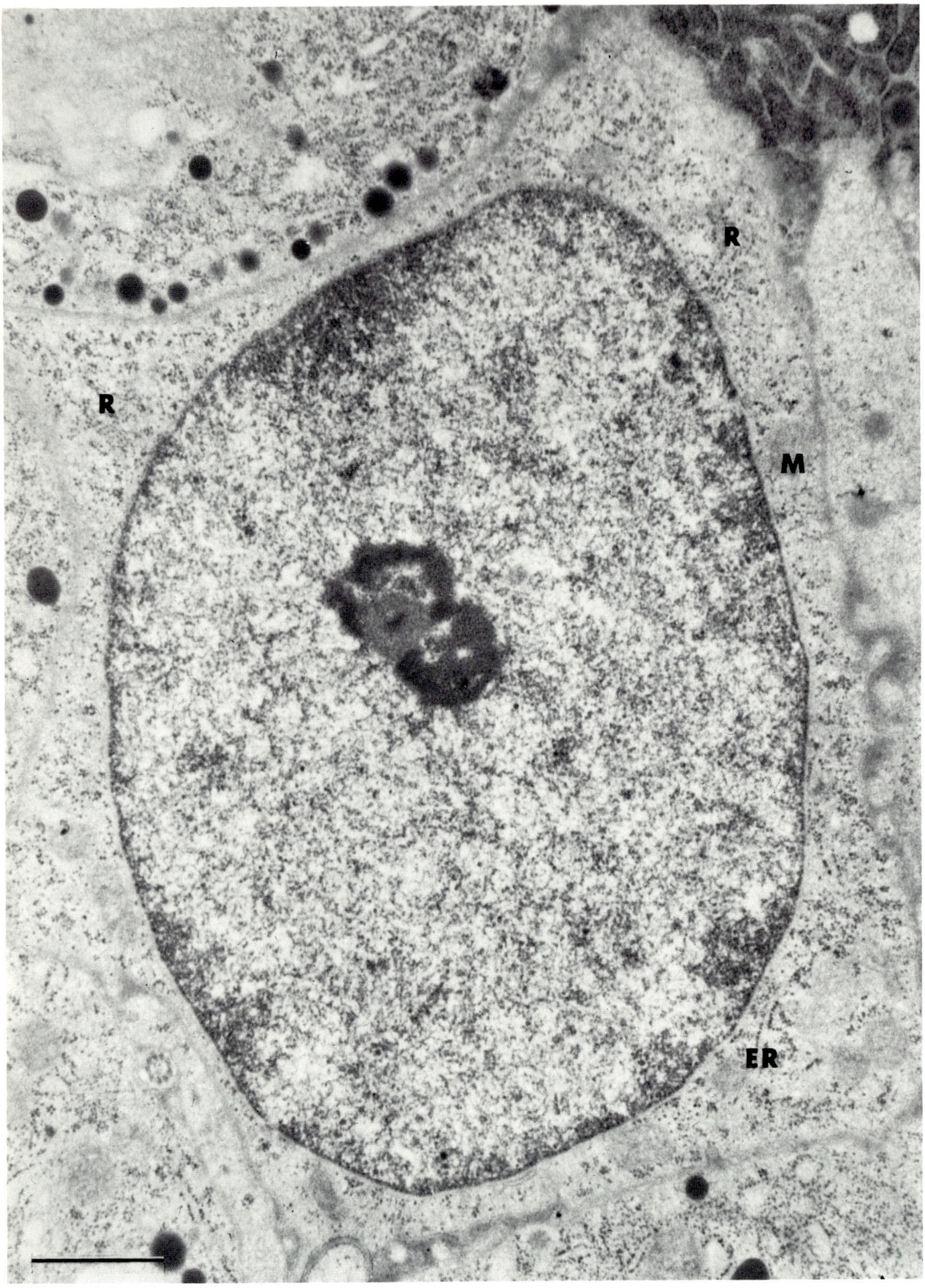

Fig. 8.1. A chromophobe from a male rat. This rather inconspicious cell has scattered areas of ribosomes (R) and several short lamellae of endoplasmic reticulum (ER) and mitochondria. × 20,000. (Glutaraldehyde–paraformaldehyde)

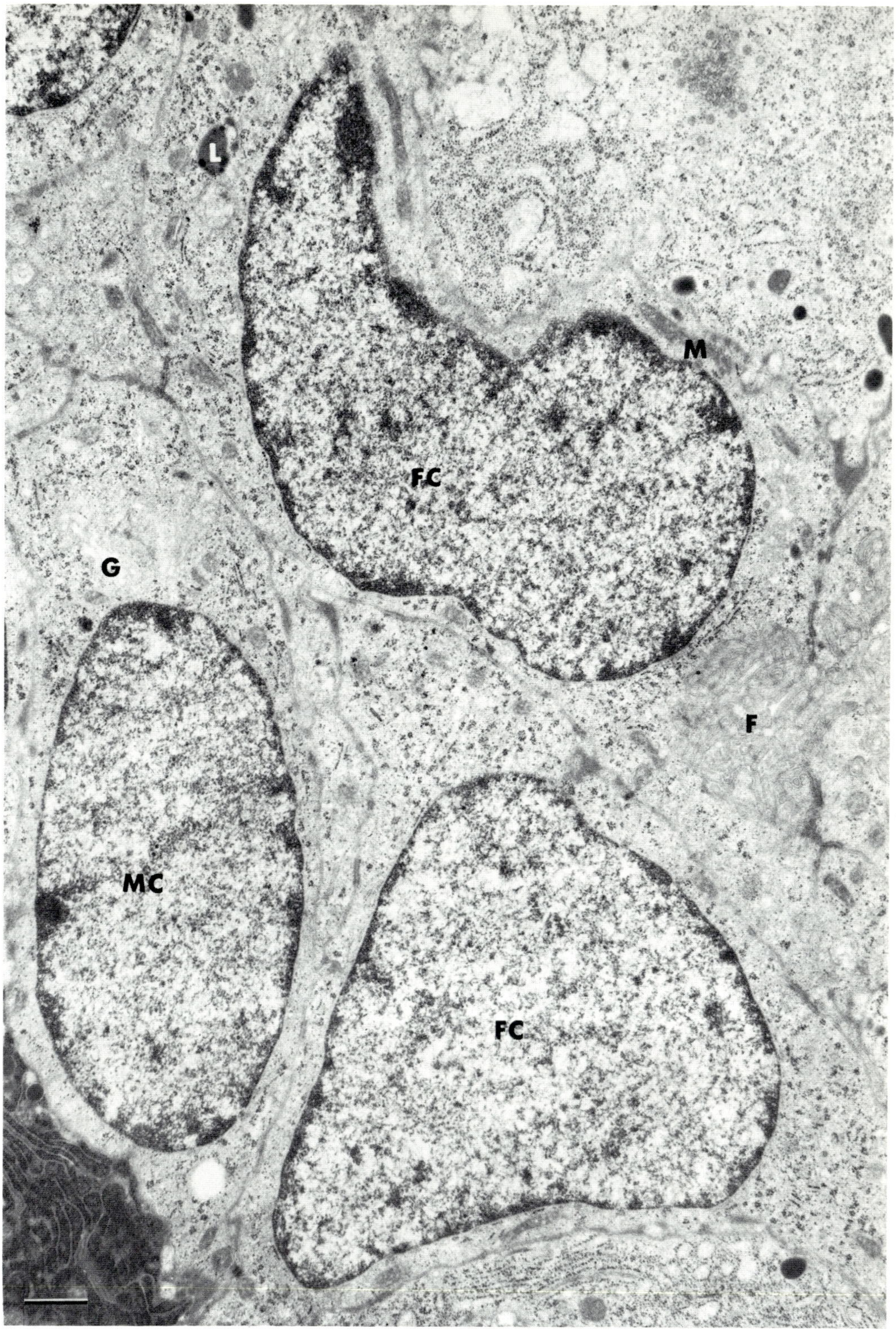

Fig. 8.2. A medium chromophobe (MC) is shown at the left-handed corner of this micrograph. It has more cytoplasm and a Golgi complex (G), otherwise it is similar to the small chromophobe of Fig. 7.1. At the right of the medium chromophobe are two follicle cells (FC) surrounding an involuted follicle (F). The follicle cells contain long, thin mitochondria (M), several Golgi complexes, and lysosomes (L). × 10,000. (Glutaraldehyde–paraformaldehyde)

pears rather inactive, and many ribosomes free in the cytoplasm. It has a few lysosomes but no granules. These cells are perhaps differentiating chromophobes, that is, small chromophobes differentiating into granulated cells. Medium chromophobes were present in the pituitary after castration or adrenalectomy, and it could be interpreted that these chromophobes were differentiating into gonadotropes and corticotropes, respectively.

Stellate Cells

As the name implies, stellate cells usually have three or four cytoplasmic projections (Fig. 8.3). These cells are medium in size, the endoplasmic reticulum consists of scattered vesicles and there are fewer ribosomes than in the medium chromophobes. A Golgi complex can often be seen and it usually appears rather hypertrophied. Dense bodies (lysosomes) are rare, and granules are not observed in these cells.

Stellate chromophobe cells have been identified by Rinehart and Farquhar (1953, 1955) in the rat, Barnes (1962) in the mouse, Salazar (1963) in the rabbit, and Cardell (1964) in the salamander. Because of the long cytoplasmic projections extending from them, Cardell (1969) reported that these cells are found throughout the pituitary and have contacts with all secretory cells in the pars distalis so that they may serve a sustentacular role. He found dense bodies in these cells but no evidence that they play a secretory role.

On the basis of ultrastructural studies of rabbit glands, Schechter (1969) postulated that the stellate cells might serve a structurally or metabolically supportive function. He noted that following Metopirone administration only the organelles of stellate cells hypertrophied. Granules were not present in these cells. On the basis of this evidence stellate chromophobes may serve a sustentacular role in the pituitary. Although in later studies Schechter (1971) suggests that stellate cells may give rise to corticotropes.

Follicle Cells

Chromophobes often appear in clusters (Fig. 8.2) surrounding an intercellular space, namely a follicle which is usually filled with colloid (Fig. 8.4). These chromophobes contain microvilli (Fig. 8.5) and cilia (Fig. 8.6) facing the lumen of the follicle and may be designated as follicle cells. Desmosomes, which are specialized areas of adhesive contact between follicle cells, were often seen. The so-called follicle surrounded by chromophobes can be more correctly represented as a pseudofollicle

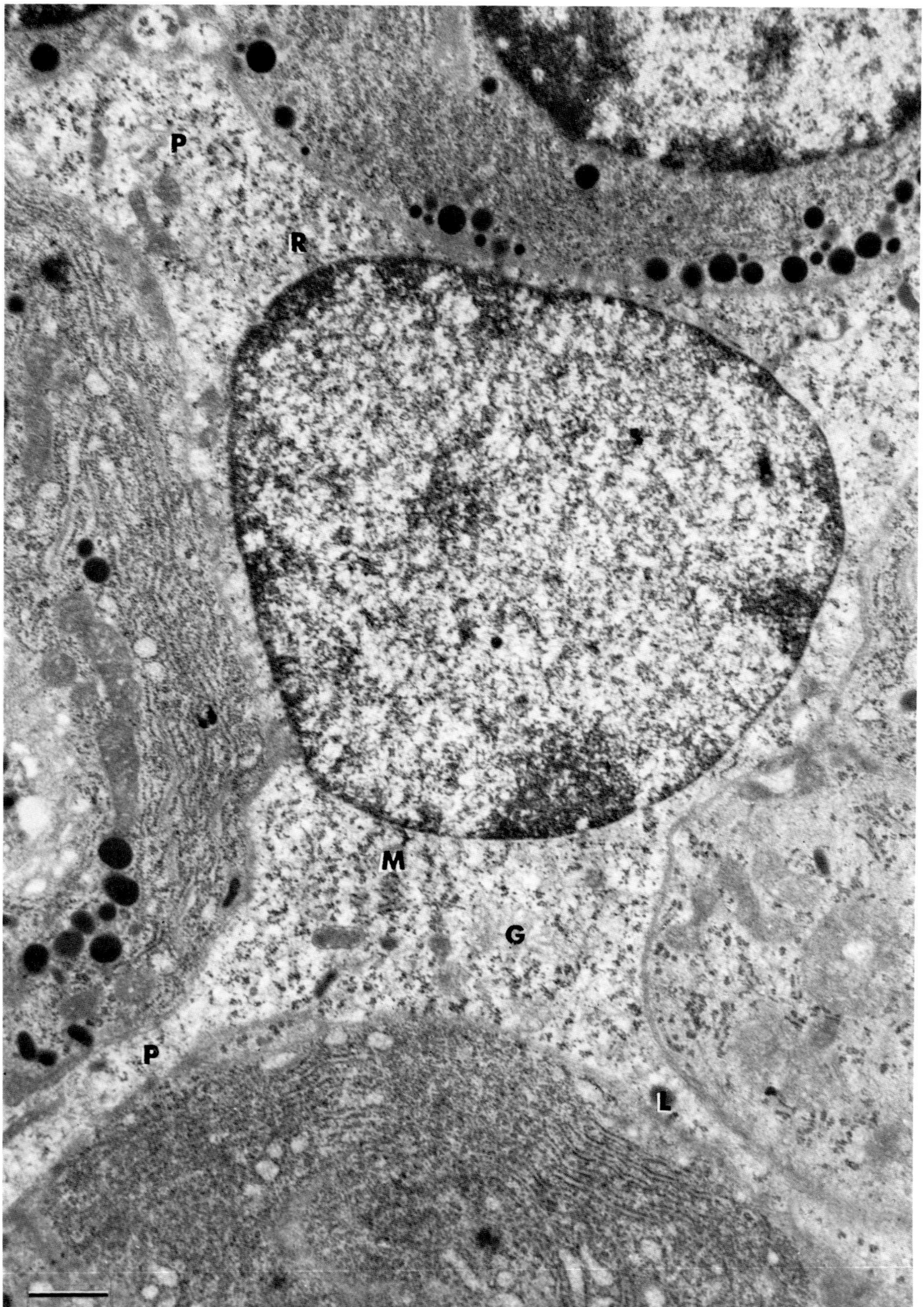

Fig. 8.3. A stellate chromophobe characterized by cytoplasmic projections (P), lysosomes (L), a Golgi complex (G), several small mitochondria (M), and an abundance of ribosomes (R). × 12,500. (Glutaraldehyde–paraformaldehyde)

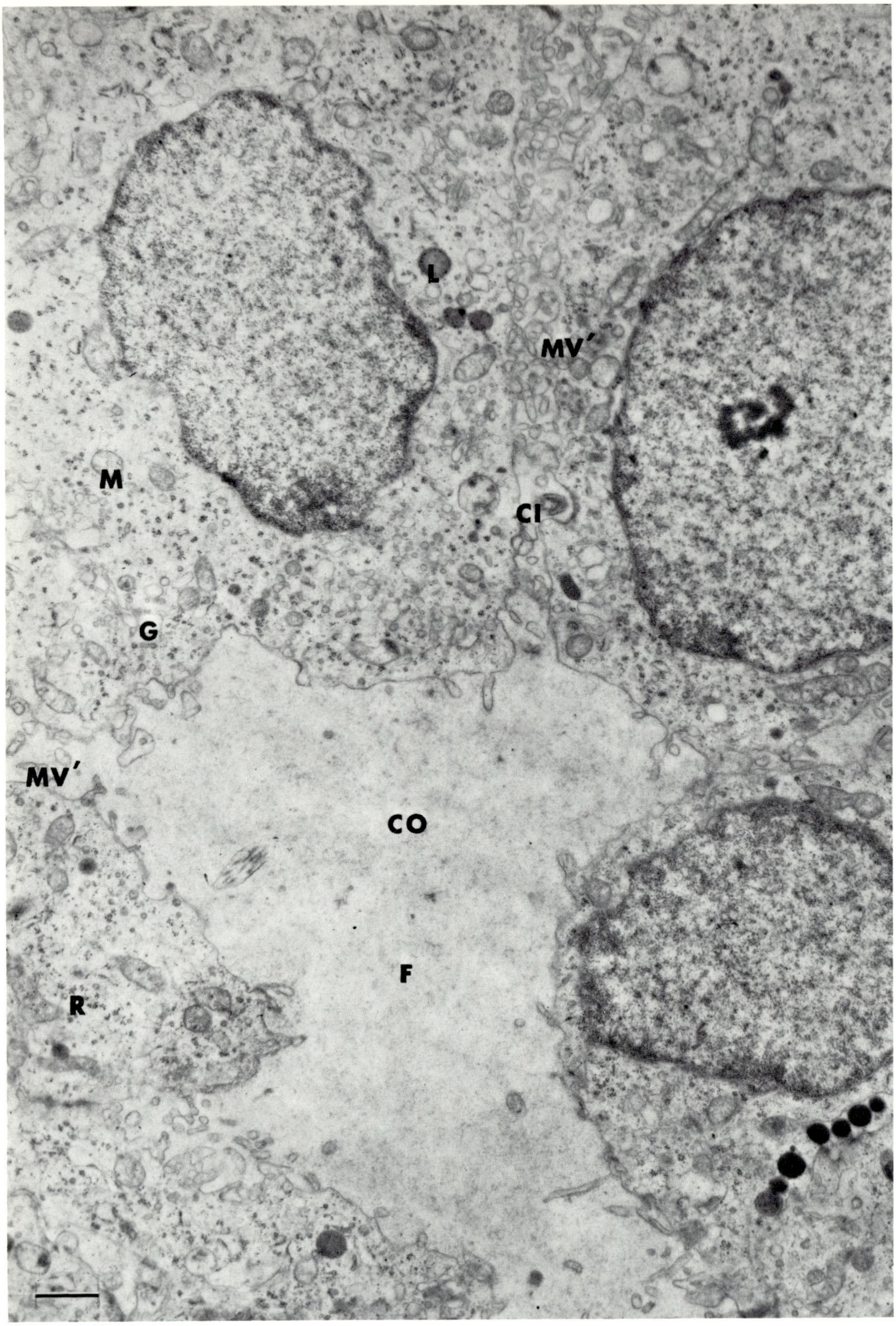

Fig. 8.4. An enlarged follicle in the pituitary 14 days after adrenalectomy. The enlarged follicle (F) containing colloid (CO) can be seen although the microvilli (MV′) are not abundant. Extensive microvilli (MV′) and a cilium (CI) are found between two follicle cells. These cells contain ribosomes (R), mitochondria (M), lysosomes (L), and Golgi complexes (G). × 10,000. (Glutaraldehyde–paraformaldehyde)

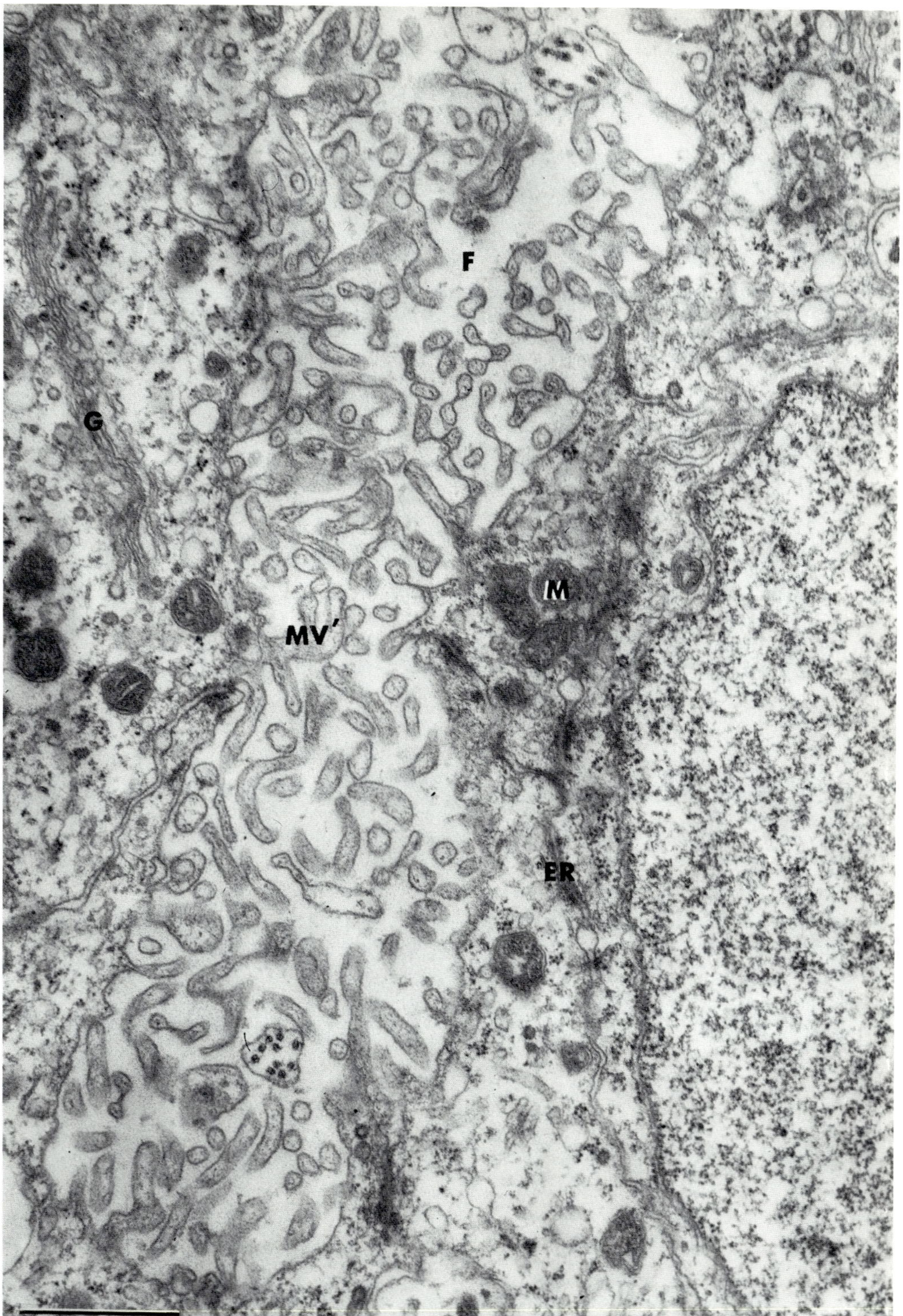

Fig. 8.5. A close-up of part of a follicle in a pituitary of an intact control rat. The follicle (F) is not enlarged but there is a well developed system of microvilli (MV) extending into the follicle from the surrounding follicle cells. An extensive Golgi complex (G), several cisternae of endoplasmic reticulum (ER), and round mitochondria (M) are present in the follicle cells. × 24,000. (Veronal acetate buffered osmic acid)

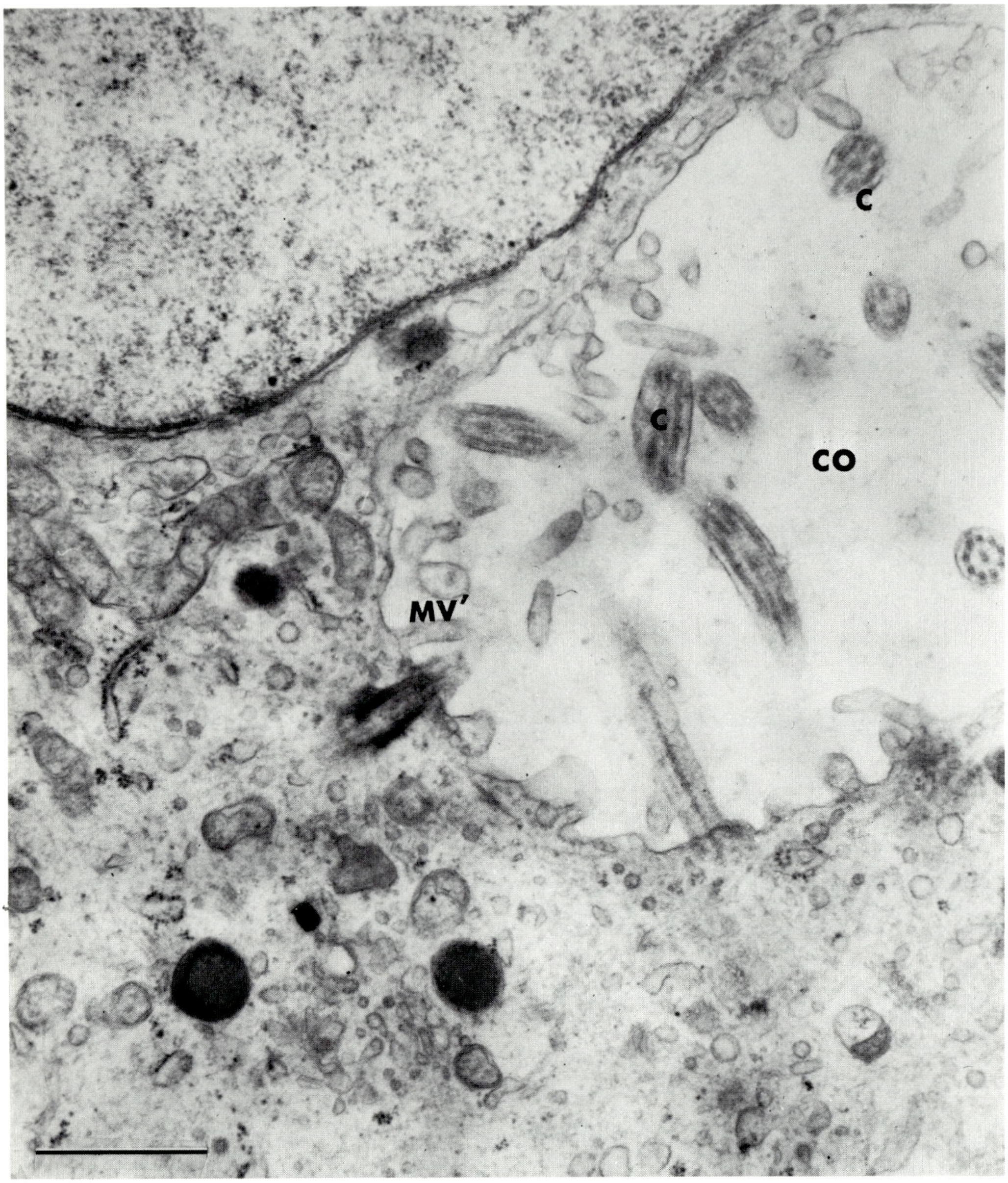

Fig. 8.6. An electron micrograph of part of a follicle at a higher magnification. Microvilli (MV′) and 9 + 2 cilia (C) extend into the colloid filled (CO) follicle from the chromophobelike follicle cells × 24,000. (Veronal acetate buffered osmic acid)

since no basement membrane surrounds the follicular wall. The numerous cilia projecting into the pseudofollicale are of the 9 + 2 fiber pattern (Fig. 8.6). The individual follicle cells contain a dense cytoplasmic matrix, a rich supply of ribosomes and a well-developed endoplasmic reticulum that is rather irregular in pattern. At least one Golgi complex was observed and it was rather hypertrophied.

Extracellular accumulation of colloid in follicles has been described in the pars distalis of the dog by Wolfe *et al.* (1933) and in other mammals by Romeis (1940). Farquhar (1957) was the first to definitively describe these follicle cells by use of electron microscopy. The follicle size and amount of colloid increased after hemi-adrenalectomy. She postulated that these cells may produce ACTH and that it is stored in the follicle. Barnes (1961) described follicle cells in the mouse pituitary gland. In the dog follicular cells are not round, medium-sized, but are stellate with long cytoplasmic projections (Kagayama, 1965). He reported that these cytoplasmic processes often reach the perivascular spaces and that these cells may serve a sustentacular role. He also observed granules up to 200 mμ in diameter in these cells. Hennigar and Paiz (1970) claimed that there are two kinds of human follicular cells, namely, chromophobe and chromophil types. They believe that the products in the colloid of the follicle may be reutilized by the follicular cell.

After adrenalectomy or castration the colloid was observed to increase in amount and the follicle to increase in size (Fig. 8.4). The exact function of these cells is not known. Because of the increase in colloid in the follicle after long-term adrenalectomy one might implicate the follicle cells in ACTH synthesis but more definitive experiments are required to test this postulation.

Pseudo Chromophobe

These cells, we believe, degranulated chromophils (see Fig. 6.6, p. 122). They are usually medium in size, contain a well-developed Golgi complex, and a dense cytoplasmic matrix with an abundance of ribosomes. The endoplasmic reticulum is usually well developed and organized in lamellar form. There is usually an abundance of rodlike mitochondria with complete cristae. These cells may contain several lysosomes and in some cases granules as well. Therefore, most of the time these pseudochromophobes are very active cells or degranulated chromophils.

In summary, little definite is known regarding the function of the chromophobes. The stem cells may, under proper physiological conditions, differentiate into chromophils. Medium chromophobes are small chromophobes that are in the process of differentiating into granulated cell types. Stellate chromophobes appear to serve a sustentacular role in the pituitary. The follicle cells and the size of the follicle change after extirpation; these cells may synthesize certain hormones.

REFERENCES

Barnes, B. G. (1961). Ciliated secretory cells in the pars distalis of the mouse hypophysis. *J. Ultrastruct. Res.* **5**, 453.

Barnes, B. G. (1962). Electron microscopic studies on the cytology of the mouse pituitary, *Endocrinology* **71**, 618.

Capen, C. C., and Koestner, A. (1967). Functional chromophobe adenomes of the canine adenohypophysis. *Pathol. Vet.* **4**, 326.

Cardell, R. R. (1964). Observations on the cell types of the salamander pituitary gland: An electron microscopic study. *J. Ultrastruct. Res.* **10**, 317.

Cardell, R. R. (1969). The ultrastructure of stellate cells in the pars distalis of the salamander pituitary glands. *Amer. J. Anat.* **126**, 429.

Ezrin, C., Swanson, H. E., Humphrey, J. G., Dawson, J. W., and Hill, F. M. (1959). The cells of the human adenohypophysis in thyroid disorders. *J. Clin. Endocrinol. Metab.* **19**, 958.

Farquhar, M. G. (1957). "Corticotrophs" of rat adenohypophysis revealed by electron microscopy. *Anat. Rec.* **127**, 291.

Fernández-Morán, M., and Luft, R. (1949). Submicroscopic cytoplasmic granules in the anterior lobe cells of the rat hypophysis as revealed by electron microscopy. *Acta Endocrinol. (Copehagen)* **2**, 199.

Hennigar, G. R., and Paiz, C. (1970). The pituitary follicular complexes. *Fed. Proc., Fed. Amer. Soc. Exp. Biol.* **29**, 362 (abstr.).

Holmes, R. L., and Mandl, A. M. (1961). A spontaneous tumor in a rat associated with adrenal hypertrophy. *J. Endocrinol.* **22**, XXIX.

Hunt, T. E. (1942). Mitotic activity in the anterior hypophysis of female rats. *Anat. Rec.* **82**, 263.

Hunt, T. E. (1951). The effect of hypophyseal extract on mitotic activity of the rat hypophysis. *Anat. Rec.* **111**, 713.

Kagayama, M. (1965). The follicular cell in the pars distalis of the dog pituitary gland: An electron microscope study. *Endocrinology* **77**, 1053.

Rasmussen, A. T. (1929a). The percentage of different types of cells in the male adult human hypophysis. *Amer. J. Pathol.* **5**, 263.

Rasmussen, A. T. (1929b). Ciliated epithelium and mucus-secreting cells in the human hypophysis. *Anat. Rec.* **41**, 273.

Rasmussen, A. T. (1933). The percentage of different types of cells in the anterior lobe of the hypophysis in the human female. *Amer. J. Pathol.* **9**, 459.

Rinehart, J. F., and Farquhar, M. G. (1953). Electron microscopic studies of the anterior pituitary gland. *J. Histochem. Cytochem.* **93**, 113.

Rinehart, J. F., and Farquhar, M. G. (1955). The fine vascular organization of the anterior pituitary gland. An electron microscopic study with histochemical correlations. *Anat. Rec.* **121**, 207.

Romeis, B. (1940) . Inkretorische organe. *In* "Handbuch der mikroskopischen Anatomie des Menschen" (W. von Möllendorf, ed.), Vol. 3, Part 3, p. 111. Springer-Verlag, Berlin and New York.

Rovit, R. L., and Berry, R. (1965). Cushing's syndrome and the hypophysis. *J. Neurosurg.* **23**, 270.

Salazar, H. (1963). The pars distalis of the female rabbit hypophysis: An electron microscopic study. *Anat. Rec.* **147**, 469.

Schechter, J. (1969). The ultrastructure of the stellate cell in the rabbit pars distalis. *Amer. J. Anat.* **126**, 477.

Schechter, J. (1971). The cytodifferentiation of the rabbit pars distalis: An electron microscope study. *Gen. Comp. Endocrinol.* **16**, 1.

Schelin, V. (1962). Chromophobe and acidophil adenomas of the human pituitary gland. *Acta Pathol. Microbiol. Scand., Suppl.* 158. 80 pp.

Severinghaus, A. E. (1933). A cytological study of the anterior pituitary of the rat with special reference to the Golgi apparatus and to cell relationship. *Anat. Rec.* **57**, 149.

Severinghaus, A. E. (1935). A suggestive correlation of cytological changes with secretory activity in the cells of the normal anterior pituitary gland. *Anat. Rec.* **61**, Suppl. 341.

Severinghaus, A. E. (1937). Cellular changes in the anterior hypophysis with special reference to its secretory activities. *Physiol. Rev.* **17**, 556.

Severinghaus, A. E. (1939). Anterior hypophyseal cytology in relation to the reproductive hormones. *In* "Sex and Internal Secretions" (E. Allen, ed.), 2nd ed., p. 45. Williams & Wilkins, Baltimore, Maryland.

Siperstein, E. R. (1963). Identification of the adrenocorticotrophin-producing cells in the rat hypophysis by autoradiography. *J. Cell Biol.* **17**, 521.

Wolfe, J. M., Cleveland, R., and Campbell, M. (1933). Cyclic histological variations in anterior hypophysis of the dog. *Z. Zellforsch. Mikrosk, Anat.* **17**, 420.

Yamada, K., Sano, M., Okumura, K., and Sakakura, K. (1960). Cellular changes in the mouse anterior pituitary from maturity to senility. *Okajimas Folia Anat. Jap.* **34**, 107.

Yoshimura, F., Harumiya, K., Ishikawa, H., and Ohtska, Y. (1969). Differentiation of isolated chromophobes into acidophils and basophils when transplanted into the hypophysiotrophic area of the hypothalamus. *Endocrinol., Jap.* **16**, 531.

Chapter 9

Pars Tuberalis

The pars tuberalis is formed from two buds derived from the structure separating the oral cavity from the aboral cavity of Rathke's pouch. These buds unite in the median line and encircle the pituitary stalk (Tilney, 1914; Herlant, 1964). In some cases, tuberal tissue remains in the region where the buds were originally formed and may appear as inclusions within the anterior lobe. This is known as the internal pars tuberalis (Dawson, 1948). In mammals the oral cavity of Rathke's pouch does not participate further in the formation of the anterior lobe but remains in the median region. The zona tuberalis forms either a solid median mass or two lateral masses.

The pars tuberalis has the richest blood supply (Greep, 1966) of the zones which form the adenohypophysis. This suggests that it has an important function in the animal. Szentágothai *et al.* (1968) reported that the pars tuberalis has an intimate relationship with the median eminence. The current viewpoints on hypothalamo-hypophyseal control mechanisms explain the necessity for a large and intimate contact zone between the hypothalamus and hypophysis. This is especially true in the lower vertebrates where there is no portal circulation. The blood supplying the hypothalamo-hypophyseal complex first reaches the pars tuberalis where the vessels form a primary plexus from which capillary loops enter the median eminence. The passage of blood is → pars tuberalis → median eminence → portal vessels → pars distalis. Furthermore, Spatz (1951) and Nowakowski (1951) postulated that blood from the median eminence would first supply the pars tuberalis before terminating in the pars distalis.

HISTOCHEMICAL STUDIES

One of the first reports on the importance of the pars tuberalis was made by Berblinger (1941), who associated gonadotropic activity with cells of the pars tuberalis. Studies by Kutas (1958) showed that TSH cells were also present.

The cells of the pars tuberalis are unique but usually have been considered to be chromophobe cells, similar to those found in the pars distalis (Herlant, 1964). Greep (1966) described the cells of the pars tuberalis as similar to those found in the anterior and intermediate lobes, although they may be smaller and poorly granulated. Fand (1965) has reported that because many of the cells of the pars tuberalis are basophils, which stain poorly with PAS, they are generally overlooked. Cells of the pars tuberalis may synthesize ACTH since Finerty and Keller (1961) found that the effects of hypophysectomy in dogs were not as severe when the pars tuberalis was left intact. They postulate that cells of this zone might synthesize ACTH.

RECENT HISTOCHEMISTRY, IMMUNOCHEMISTRY, AND ELECTRON MICROSCOPY OF THE PARS TUBERALIS

Some cells of the pars tuberalis in man react with human chorionic gonadotropin (HCG) antibodies, and Midgley (1963) concluded that certain of the cells must contain LH. This work confirmed the earlier experimental conclusions by Hayashida (1962). Reichlin (1963) reported that TSH and ACTH as well as gonadotropins were associated with extracts of pars tuberalis. In studies by Fand and Thorell (1962) TSH, FSH, LH and possibly ACTH cells were shown by histochemical methods to be found in this zone. Furtherfore, Fand (1965) performed studies which suggest α-glycerophosphate dehydrogenase (αGPD) in the human pituitary is associated with the synthesis and release of TSH. This enzyme was found in cells of the pars tuberalis, suggesting that the cells may also produce TSH.

More evidence suggesting that LH is present in the pars tuberalis of human and bovine pituitary glands was provided by Midgley (1966) and Dubois (1970). Using immunohistochemical methods they found a marked concentration of LH gonadotropes in the pars tuberalis and the continuation of these cell nests over the superior surface of the anterior lobe. In contrast, LH cells in the pars distalis did not appear in groups but were found singly and did not stain to the same extent as those in the pars tuberalis.

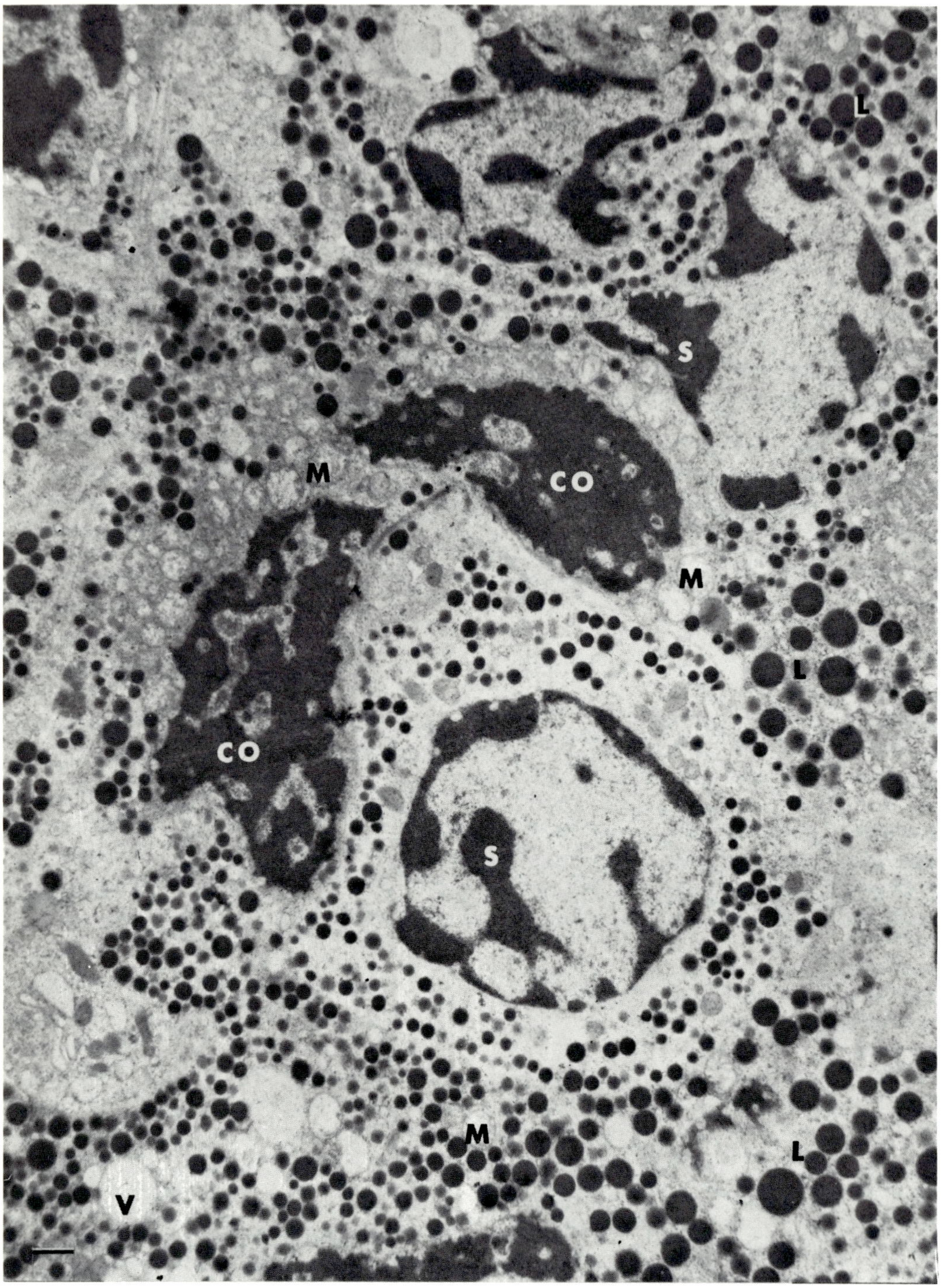

Fig. 9.1. A section of the pars tuberalis. Cells with large granules comparable to somatotropes (S) and luteotropes (L) of the pars distalis are shown. Round mitochondria (M) and vacuoles (V) are evident. Areas of colloid (CO) can be seen among the cells. × 6000. (Glutaraldehyde–paraformaldehyde)

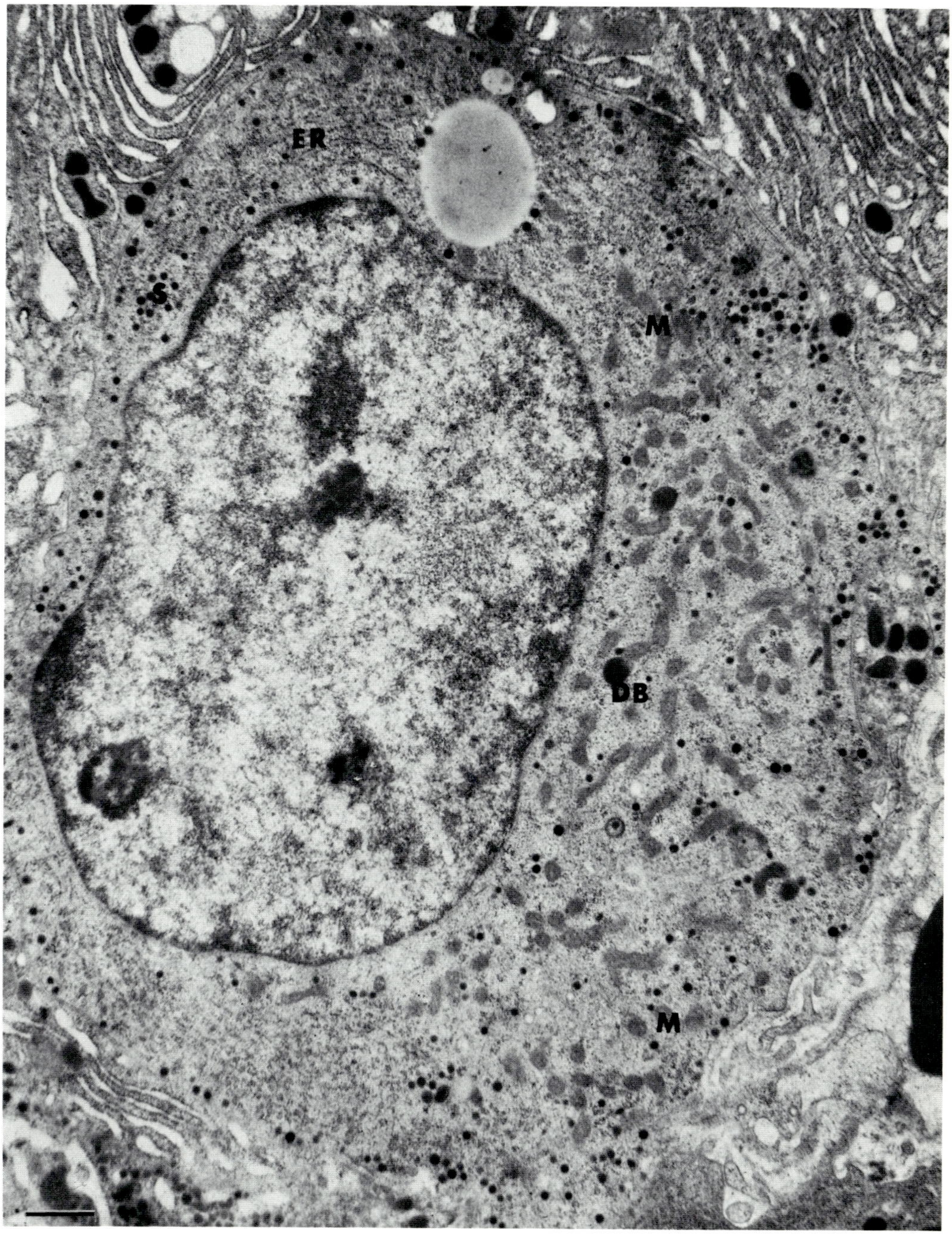

Fig. 9.2. Pars tuberalis. A cell with small granules (S) of up to 150 mμ maximum diameter is shown. It has some similar features of a corticotrope or thyrotrope from the pars distalis. Round and short rod mitochondria (M), dense bodies (DB) (presumably lysosomes), and several short lamellae of rough endoplasmic reticulum (ER) are present. $\times$ 10,000. (Glutaraldehyde?paramormaldehyde)

Legait (1969) reported data which have substantiated the findings of Midgley (1966) and Dubois (1970). He injected prepubertal rats with homogenates of pars tuberalis and found stimulation of interstitial cells of the testes and luteal tissue of the ovary. On the basis of these results he postulated that the pars tuberalis produces LH. One might question whether the tuberalis extracts were contaminated with infundibular tissue. If there was such contamination, neurosecretory factors might stimulate the pituitary gland of the prepubertal rats to secrete the LH. Legait and Contet (1969) did histophysiological experiments on the pars tuberalis of the rat and found that it is composed of two parts; in one, gonadotropes predominate, and in the other, acidophils and chromophobes are more prevalent. They also found that after castration or during gestation these gonadotropes hypertrophy and the volume of the pars tuberalis increases.

In our electron microscopy studies, we found that virtually all cell types of the pars anterior were present in the pars tuberalis. Cells with large granules (Fig. 9.1), presumably STH and LTH, and gonadotropelike cells (Fig. 9.2), especially LH cells, appear to predominate. Thyrotropes and chromophobes as identified in the pars distalis are few, if any, in number. Invariably, a part of the infundibulum, intimately associated with the pars tuberalis, is also observed.

Stutinsky *et al.* (1964) conducted ultrastructural studies of the remaining pars tuberalis after hypophysectomy in rats and report the presence of FSH and ACTH cells. Dubois and Cohere (1970) also conducted electron microscopic studies of the pars tuberalis of the cow. They observed somatotropes, LH gonadotropes, and mammotropes in the bovine pars tuberalis.

Summary

From the review of the literature and the ultrastructural findings it appears that some of the cells of the pars tuberalis are similar to those found in the pars distalis. Perhaps there are even cells producing other tropic factors not hitherto identified in this region of the pituitary. It appears that the pars tuberalis might serve an important function in governing the gonads and even producing ACTH but these postulations remain to be established.

REFERENCES

Berblinger, W. (1941). Ist die Pars Tuberalis der Hypophyse Gonadotrope Wirksam?. *Endokrinologie* **23**, 251.

Dawson, A. B. (1948). The relationship of pars tuberalis to pars distalis in the hypophysis of the Rhesus monkey. *Anat. Rec.* **102**, 103.

Dubois, M. P. (1970). Cytologie de l'hypophyse de bovins separation des cellules somatotropes et des cellules à prolactine par immunofluorescence identification des cellules LH dans la pars tuberalis et la par intermedia. *C. R. Ass. Anat., Sofia* p. 139.

Dubois, M. P., and Cohere, G. (1970). Cytologie ultrastructurale du lobe antérieur et de la pars tuberalis de l'hypophyse de bovins. *C. R. Ass. Anat., Sofia* p. 147.

Fand, S. B. (1965). Demonstration of α-glycerophosphate dehydrogenase in the human pituitary gland with special reference to the cells of the pars tuberalis. *Endocrinology* **77**, 320.

Fand, S. B., and Thorell, B. (1962). Spectrophotometry of the periodic acid–Schiff reaction with pituitary hormones *in vitro* and in histological sections. *J. Cell Biol.* **13**, 239.

Finerty, J. C., and Keller, A. D. (1961). Regional localization of adrenocorticotrophic cells in the anterior hypophysis of the dogs. *Anat. Rec.* **139**, 228.

Greep, R. O. (1966): *In* "Histology," p. 743 2nd ed. McGraw-Hill (Blakiston), New York.

Hayashida, T. (1962). Immunological Studies with pituitary lactogenic hormone (prolactin) . *Ciba Found. Colloq. Endocrinol.* **14**, 326.

Herlant, M. (1964). The cells of the adenohypophysis and their functional significance. *In. Rev. Cytol.* **17**, 299.

Kutas, J. (1958). Die Thyrotrope Activitätverschiedenen Teile der Hypophyse. *Acta Biol. (Budapest)* **9**, 99.

Legait, H. (1969). Recherches experimentales sur le lobe tuberal de l'hypophyse chez le rat impubère. *C. R. Soc. Biol.* **163**, 489.

Legait, H., and Contet, J. L. (1969). Etude histophysiologique de la pars tuberalis de l'hypophyse chez le rat. *C. R. Soc. Biol.* **163**, 489.

Midgley, A. R., Jr. (1963). Immunofluorescent localization of human pituitary luteinizing hormone. *Exp. Cell Res.* **32**, 606.

Midgley A. R., Jr. (1966). Human pituitary luteinizing hormone: An immunohistochemical study. *J. Histochem. Cytochem.* **14**, 159.

Nowakowski, H. (1951). Infundibulum und Tuber Cinareum der Katze. *Dent. Z. Nervenheilk.* **165**, 201.

Reichlin, S. (1963). Neuroendocrinology. *N. Engl. J. Med.* **269**, 1182.

Spatz, H. (1951). Neues Über die Verknupfung von Hypophyse and Hypothalamus. *Acta Neuroveg.* **3**, 1.

Stutinsky, F., Porte, A., and Stoeckel, M. E. (1964). Sur les modifications ultrastructurales de la pars tuberalis du rat apres hypophysectomie. *C. R. Acad. Sci.* **259**, 1765.

Szentágothai, J., Flerko, B., Mess, B., and Halasz, B. (1968). "Hypothalamic Control of the Anterior Pituitary," p. 78. Akadémiai Kiadó, Budapest.

Tilney, F. (1914). An analysis of the juxtaneural epithelial portion of the hypophysis cerebri with an embryological and histological account of a hitherto undescribed part of the organ. *Int. Monatsschr. Anat. Physiol.* **30**, 258.

Chapter 10

Pars Intermedia

In 1919 Atwell reported that extracts from mammalian pituitary glands darkened the skin of hypophysectomized frogs. This active principle was considered a hormone and has been referred to as intermedin, melanophore expanding principle or melanophore hormone. Since the melanocyte is termed the melanin-forming cell in animals (Gordon, 1953) and because mammalian as well as amphibian melanocytes are affected by it, this active principle has been designated the melanocyte-stimulating hormone (MSH).

Although it is known that MSH is associated with the pars intermedia and its function has been known in amphibians and fishes for many years, little is known about its function in mammals. Itano (1936) and Bailif (1938) reported morphological studies of the pars intermedia but they could not correlate morphology with function. Much of the early literature on the pars intermediate has been reviewed by Legait (1964) and Wingstrand (1966).

The pars intermedia is composed of three zones: *(1)* a rim of one to three layers of chromophobe type cells lining the residual hypophyseal cleft (these cells are called cleft cells); *(2)* light and dark chromophil cells of the central region of the pars intermedia (about twelve layers of cells); and *(3)* dark cells facing the pars nervosa and axonic end bulbs containing neurosecretory granules most likely originating from the pars nervosa.

HYPOPHYSEAL CLEFT

The pituitary cleft or residual lumen of Rathke's pouch is a space between the pars distalis and the pars intermedia lined by ciliated chromophobelike cells, cleft cells. The residual Rathke's pouch was first observed by Bryant (1916) who reported that the lumen was lined with ciliated cells and contained colloid. Rasmussen (1929) observed gobletlike cells surrounding the cleft in the rat pituitary. Ferrer (1956) conducted extensive studies of the hypophyseal cleft and noted an increase in the size of the cleft and amount of colloid after adrenalectomy and castration. Vanha-Perttula (1966) found after estrogen, testosterone, or cortisone treatment that the cleft in the rat is reduced in size, is difficult to observe, and is usually devoid of colloid.

In our studies hypophyseal cleft cells (Fig. 10.1) are medium in size and similar to follicle cells. The endoplasmic reticulum is usually vesicular and rather poorly developed. A Golgi complex is often present. Mitochondria appear as round or short rods with discontinuous cristae. There are numerous free ribosomes in the cytoplasm, but the cytoplasmic matrix is not very dense. Cleft cells usually are attached to one another by desmosomes (Fig. 10.1). These cells often lack basement membranes and, although they often contain dense bodies (lysosomes), granules are rarely seen. Cells of the pars distalis lining the cleft are similar to the cleft cells of the pars intermedia. The only difference is that some basophilic or acidophilic cells are sometimes found between the cleft cells on the pars distalis side. The free surface of the cleft cells contains microvilli (Fig. 10.2) of varying sizes and cilia with the 9 + 2 fiber pattern.

Not only do cells lining the cleft contain cilia but cells of the anterior pituitary, especially acidophils and gonadotropes, have also been described with one or more cilia (Fig. 10.3). Barnes (1961), Millhouse (1967), Wheatley (1967), and Dubois and Girod (1970) have observed cilia in cells of the anterior pituitary glands of different species. These cells have cilia of the 9 + 0 pattern, while cleft cells have cilia of the 9 + 2 pattern. The 9 + 2 cilia are found also in the trachea, oviduct, and other organs and may have a motile or sensory function. Kurosumi *et al.* (1961) and Carpenter (1962) have observed movement of cilia in cleft cells. Carpenter (1971) reported that cilia in the cleft cells are especially prevalent in the 2-month-old rat as compared to other ages. The 9 + 0 cilia are usually thought to be vestigial or degenerate. However, Wheatley (1967) postulated that the 9 + 0 type may play a role in mitosis, serve a motile function, or have a sensory (chemoreceptive) function.

The Source of the Colloid Material in the Cleft

Lewis *et al.* (1937) believed intermedin (MSH) is stored in the cleft. Iturriza and Koch (1964) postulated that the colloid was an intracellular storage form of MSH bound to glycoprotein. The colloid may be due to the destruction of acidophils or basophils, transudation of plasma from

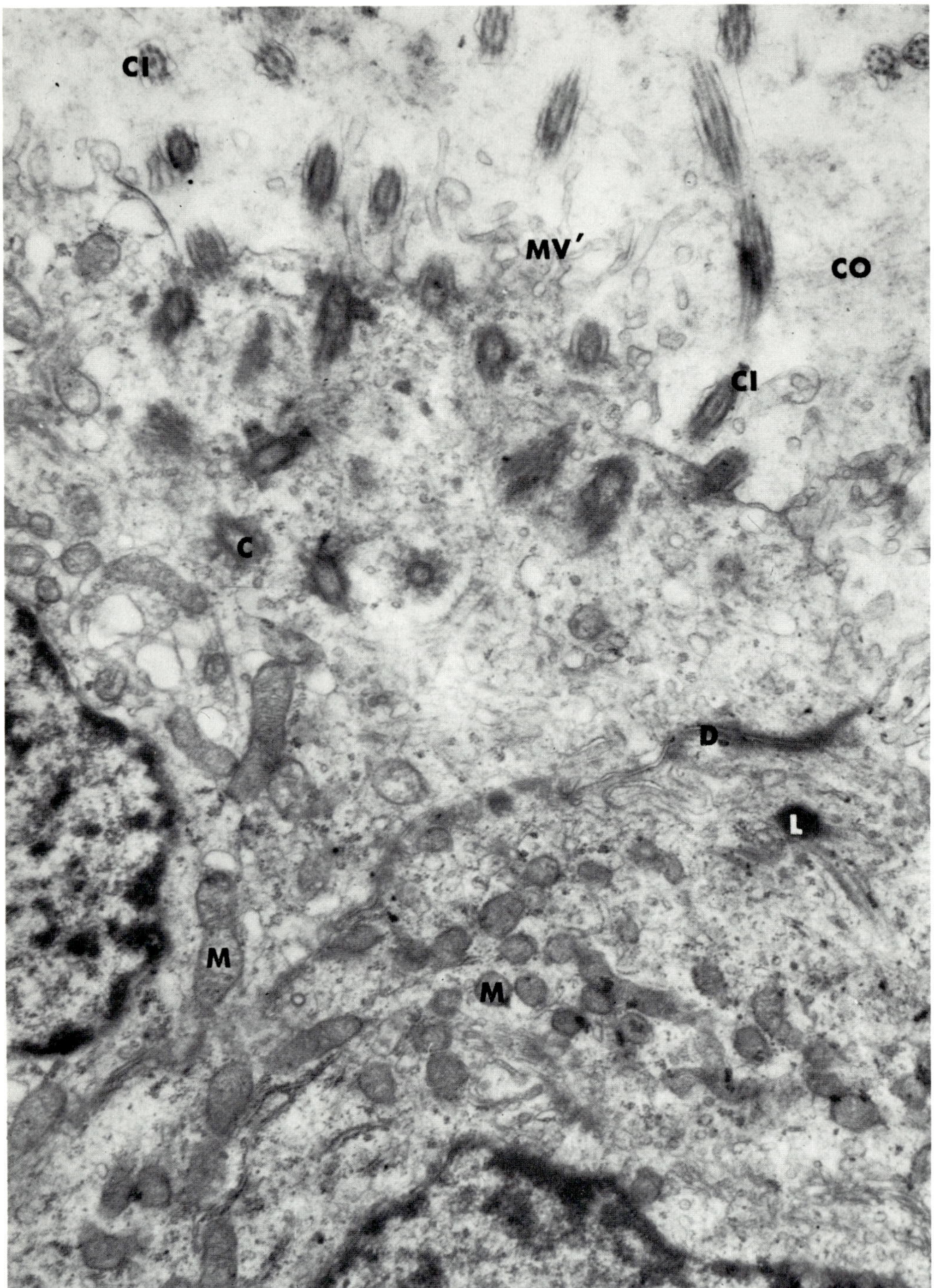

Fig. 10.1. A portion of the hypophyseal cleft microvilli (MV′) and centrioles (C) with cilia (CI) extending from them are sent into the cleft from the cleft cells. Colloid (CO) appears to fill the cleft. A lysosome (L), mitochondria (M), and a desmosome (D) are also present. × 18,000. (Glutaraldehyde–paraformaldehyde)

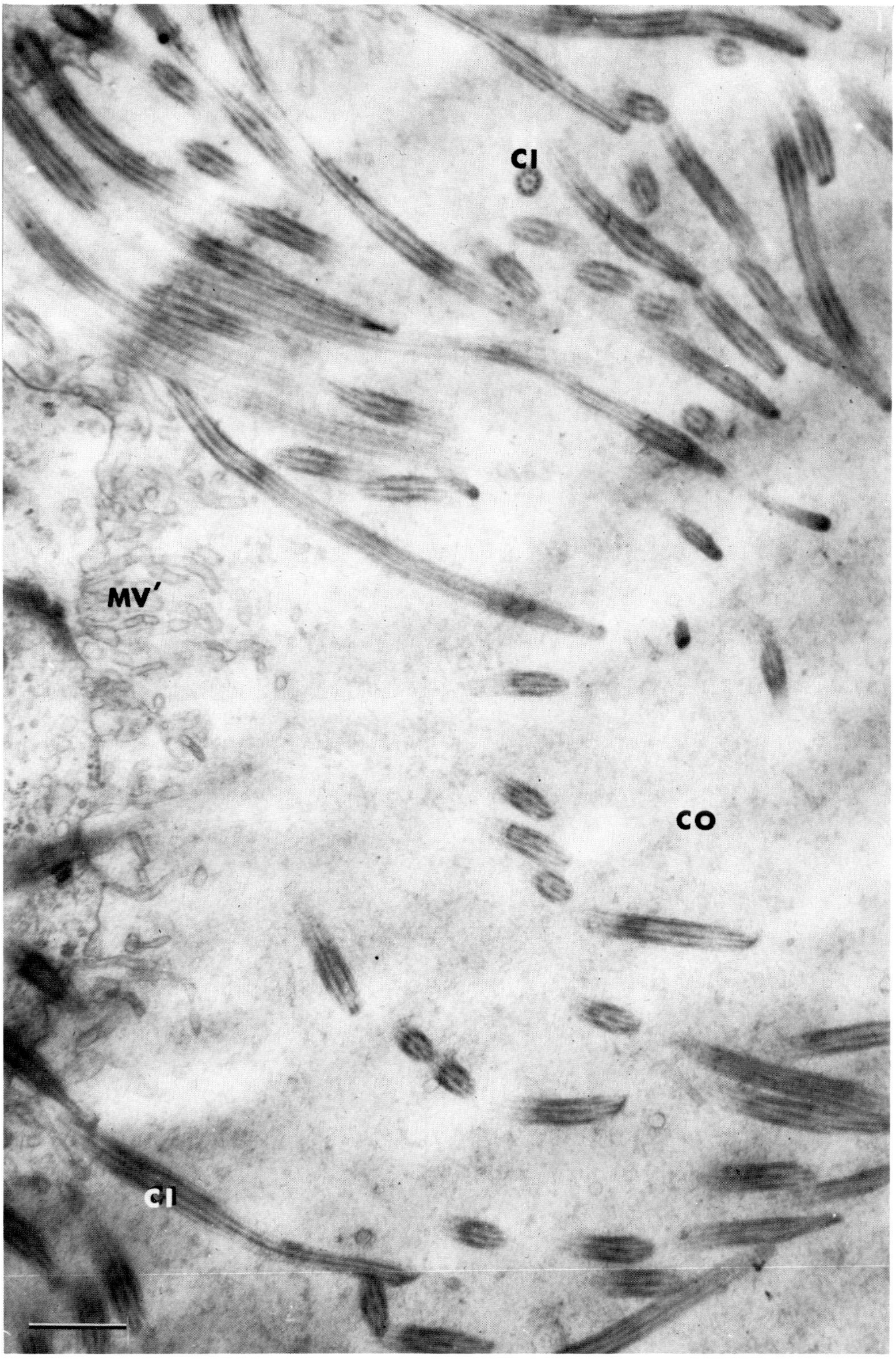

Fig. 10.2. A section through the hypophyseal cleft 14 days after adrenalectomy. The cleft is greatly enlarged and contains colloid (CO). The cleft cells show an increase in cilia (CI) and microvilli (MV′). × 15,000. (Glutaraldehyde–paraformaldehyde)

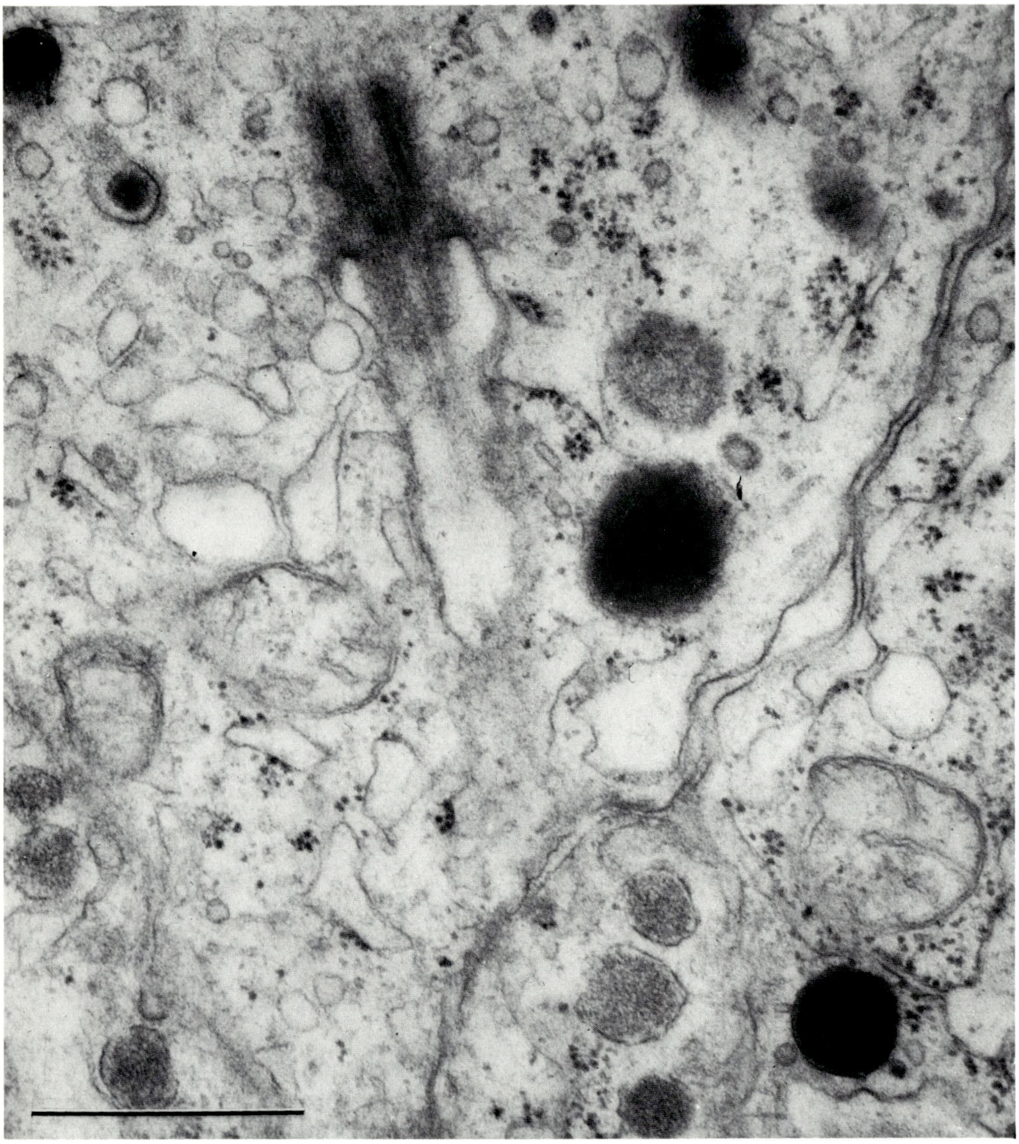

Fig. 10.3. A cilium extending from a somatotrope is shown in this micrograph. × 37,500. (Glutaraldehyde-paraformaldehyde)

the capillaries in the anterior wall (pars distalis side) of the cleft, or to secretions from the cleft epithelium (Ferrer, 1956). Vanha-Perttula (1966) observed broken cells from the pars distalis in the cleft colloid, which stained as though it were glycoprotein. Electron microscopic studies of Paiz (1970) showed holocrine secretion of the cleft cells into the lumen of human pituitary glands. Vanha-Pertulla and Arstila (1970) reported that there are high levels of oxidative enzymes in the cleft cells and postulated that these cells serve in the transfer or degradation of the colloid of the cleft.

Association of the colloid with neurosecretory material was based on Gomori positive staining by Kurosumi *et al.* (1961, 1962). They postulated that the colloid is a neurosecretory substance released by axons terminating in the pars intermedia and then transported to intercellular spaces and hence into the lumen. They suggested that secretions from the pars intermedia also are present in the cleft. Since few granules are present in the dark cells of the intermedia, these cells may directly secrete a colloid which is stored in the cleft.

In summary, since the cleft and colloid increase after adrenalectomy, and ACTH and CRF also increases, it is reasonable to postulate that the colloid contains ACTH (from the pars distalis or pars intermedia), CRF or other neurosecretory substances. More research is required to test and confirm this hypothesis.

CELLS OF THE CENTRAL REGION OF THE PARS INTERMEDIA

Trautman (1911) described the light and dark cells in the pars intermedia. Histochemical studies by Romeis (1940) in the human being, and Kurosumi *et al.* (1962) in rats have verified this early description. Early electron microscopic studies by Kurosumi *et al.* (1961) and Ziegler (1963) reported two different cell types in the intermediate lobe. However, Duchen (1962) and Raftery (1969) questioned whether there are two cells or only one in different stages of differentiation. More recently Howe and Maxwell (1968) and Kobayashi (1965) have made more complete ultrastructural studies of the pars intermedia of the rat pituitary.

The cell type that predominates is the light type I cell. This cell was so designated because it did not stain intensely with PAS (Purves, 1961) as compared with the PAS-positive type II dark cell. Because MSH is not a glycoprotein, Purves has postulated that a carrier substance in the secretory granule may be responsible for the PAS-positive staining of the latter cell. Otherwise the light cell type was so designated because of its clear

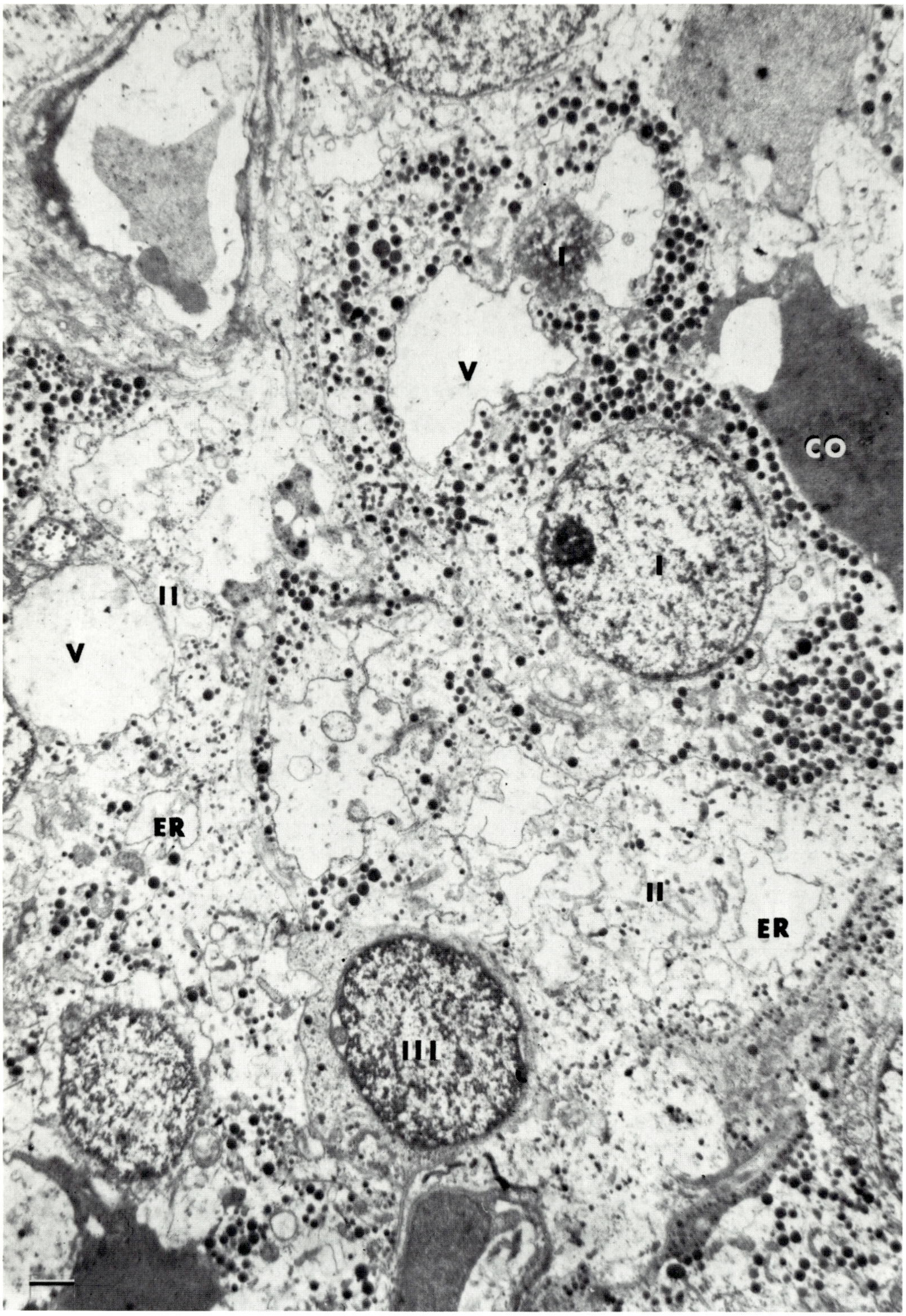

Fig. 10.4. A typical section of pars intermedia. Parts of two type I light cells (I) and type II dark cells (II) are shown. A small cell (III) with scant dense cytoplasm is also present. Large vacuoles (V) or dilated endoplasmic reticulum (ER) are found within some cells and areas of colloid (CO) among cells are present. × 6000. (Glutaraldehyde–paraformaldehyde)

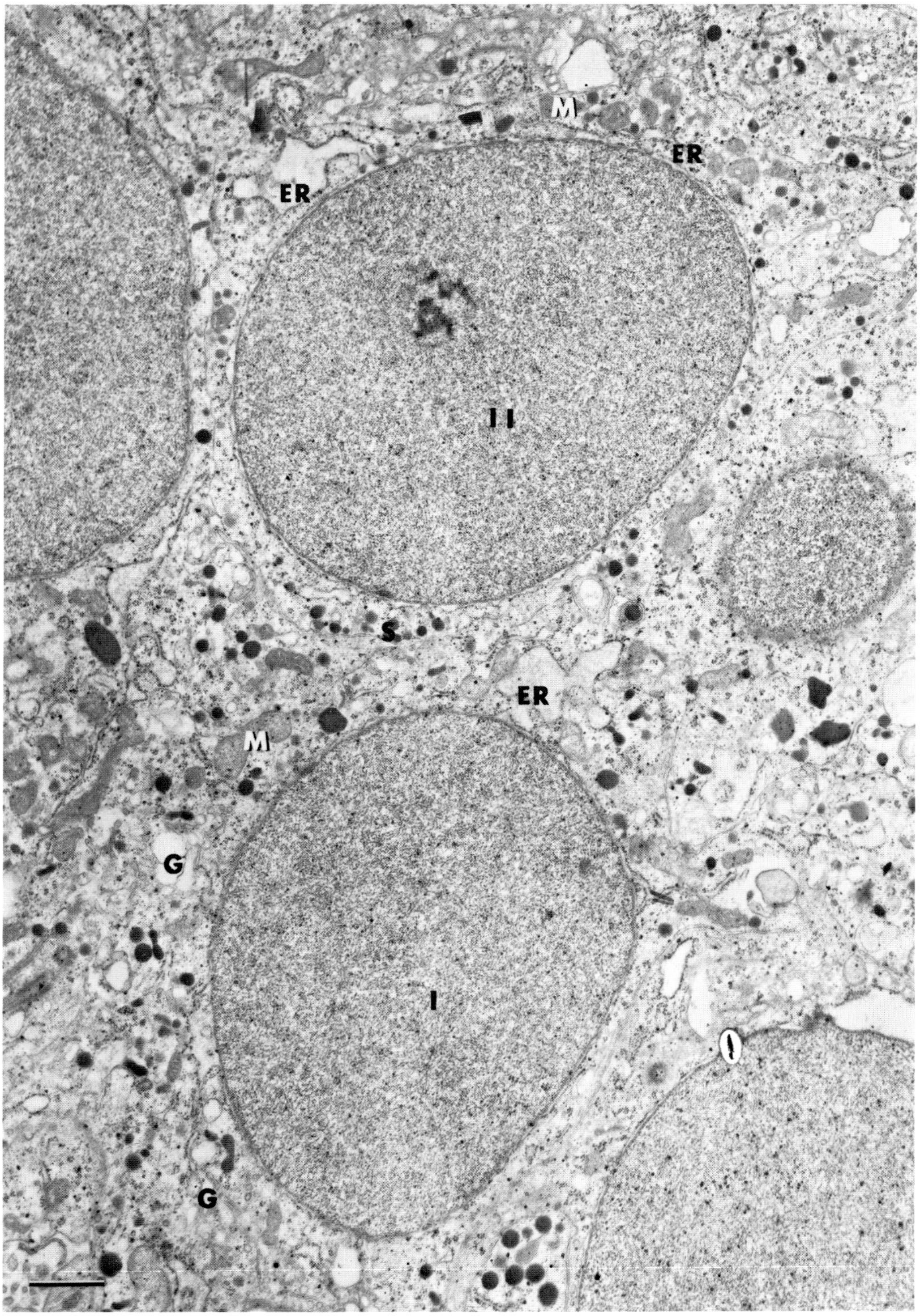

Fig. 10.5. Type I cell (I) contains rod-like mitochondria (M) and several small Golgi complexes (G). The type II cell (II) has little cytoplasm, small granules (S), and round or oblong mitochondria (M). In both cells several dilated cisternae of endoplasmic reticulum (ER) are present. $\times$ 10,000 (Glutaraldehyde–paraformaldehyde)

cytoplasmic matrix and abundance of vacuoles (Kobayashi, 1964). With glutaraldehyde fixation the cytoplasmic matrix is darker.

Type I cells are more abundant, comprising perhaps 70% of the pars intermedia (Fig. 10.4). These cells are medium sized, rather cuboidal, and have a large round nucleus. Mitochondria are rather round or short rods with incomplete cristae, a dense matrix, and are usually found surrounding a rather inconspicuous Golgi complex (Figure 10.5). Vesicular or vacuolar rough endoplasmic reticulum and free ribosomes are scattered throughout the cytoplasm. These ribosomes are not usually in rosette shape complexes as are found in cells of the pars distalis. Vacuoles of different sizes are present in certain cells as are lysosomes. Granules are usually aggregated at one of the cell poles and around the cell membrane. They are 300–350 mμ in diameter (Fig. 10.4). The cells nearer the pars nervosa are better granulated. Although the granules of type I cells were usually quite regular in size, several cells were observed to have large round granules, large granules of irregular shapes (similar to LTH granules), and smaller ones approximately 200 mμ in diameter. These cells predominate at the junction of the pars distalis.

Type II dark cells are few in number. They contain granules 200 mμ in diameter and are found throughout (Fig. 10.4) this rather angular or stellate shaped cell. There is often a lack of vacuolated areas but the cytoplasmic matrix is not as dense with glutaraldehyde fixation, as is the type I cell. Some cells contain large dilated areas of endoplasmic reticulum (Fig. 10.4) studded with ribosomes. Mitochondria tend to be rodlike (Fig. 10.5) with complete cristae. Very few Golgi complexes are present in these cells.

Another type of cell is observed very infrequently. This cell may be likened to the pars distalis small chromophobe. It has a large nucleus and very little dense cytoplasm (Fig. 10.4). Free ribosomes and several granules or dense bodies 250–300 mμ in diameter are present. This cell may correspond to the rhomboid cell of Kobayashi (1965). Whether it is a chromophobe-type cell or a cell that is differentiating into one of the other types remains to be determined.

In general our findings show that there are large vacuolated areas and areas filled with colloid (Fig. 10.4) within and among the cells. Several capillaries were observed which is contrary to the findings of Howe and Maxwell (1968). However, the blood supply of this region is very poor.

The Pars Intermedia Facing the Pars Nervosa

As early as 1894 Ramón y Cajal reported that nerve fibers extend from the pars nervosa to the pars intermedia. Sections of the intermediate lobe near the pars nervosa (adjoining the neural lobe) showed many nerve axons and type I cells. Unmyelinated nerve fiber end bulbs were found containing small vesicles but others were devoid of neurosecretory material.

Cells of Pars Intermedia after Adrenalectomy

Kobayashi (1965) observed ultrastructural changes in cells of the pars intermedia after formalin-induced stress and adrenalectomy. He found that granulation greatly increased in light type I cells. These granules were similar to neurosecretory vesicles of axons and he suggested that there is a transition from 350 mμ granules to 200 mμ vesicles in the type I cells.

Using autoradiography, Gosbee *et al.* (1970) have shown a functional relationship between the pars intermedia and ACTH secretion. They found that after adrenalectomy the small stellate type II cell showed an increased activity, granulation, and an increase in colloid. Under these conditions MSH concentration in the pituitary and plasma was not increased but ACTH was increased. The implication is that type II cells, especially after adrenalectomy, are involved in the synthesis of ACTH. Using immunoglobulin–peroxidase methods, Phifer and Spicer (1970) reported that ACTH occurs in the pars intermedia of many species of animals (including the rat). Porte *et al.* (1971) have found corticotropes in the rat pars intermedia and Kraicer *et al.* (1971) reported a greater concentration of ACTH in the intermediate lobe than in the anterior lobe of the rat.

CELLS OF THE PARS INTERMEDIA AFTER PITUITARY IS AUTOGRAFTED TO THE KIDNEY CAPSULE

The work of Taleisnik *et al.* (1966) not only proved the existence of a MSH-releasing factor (MSH-RF) but traced its origin to the paraventricular nucleus of the hypothalamus, Kastin and Schally (1966, 1967) and Kastin *et al.* (1969) reported that exogenous MSH increased MSH storage in the pituitary gland and that the pineal melatonin decreased MSH. They also reported that there is an MSH inhibitory factor in the hypothalamus (MSH-IF).

Rust and Meyer (1968) placed pituitary autografts under the kidney capsule of weasels. After the grafts were placed, the animals molted and their pelage was a dark brown, darker than the intact control animals. This pelage continued to be dark brown even when the animals were ex-

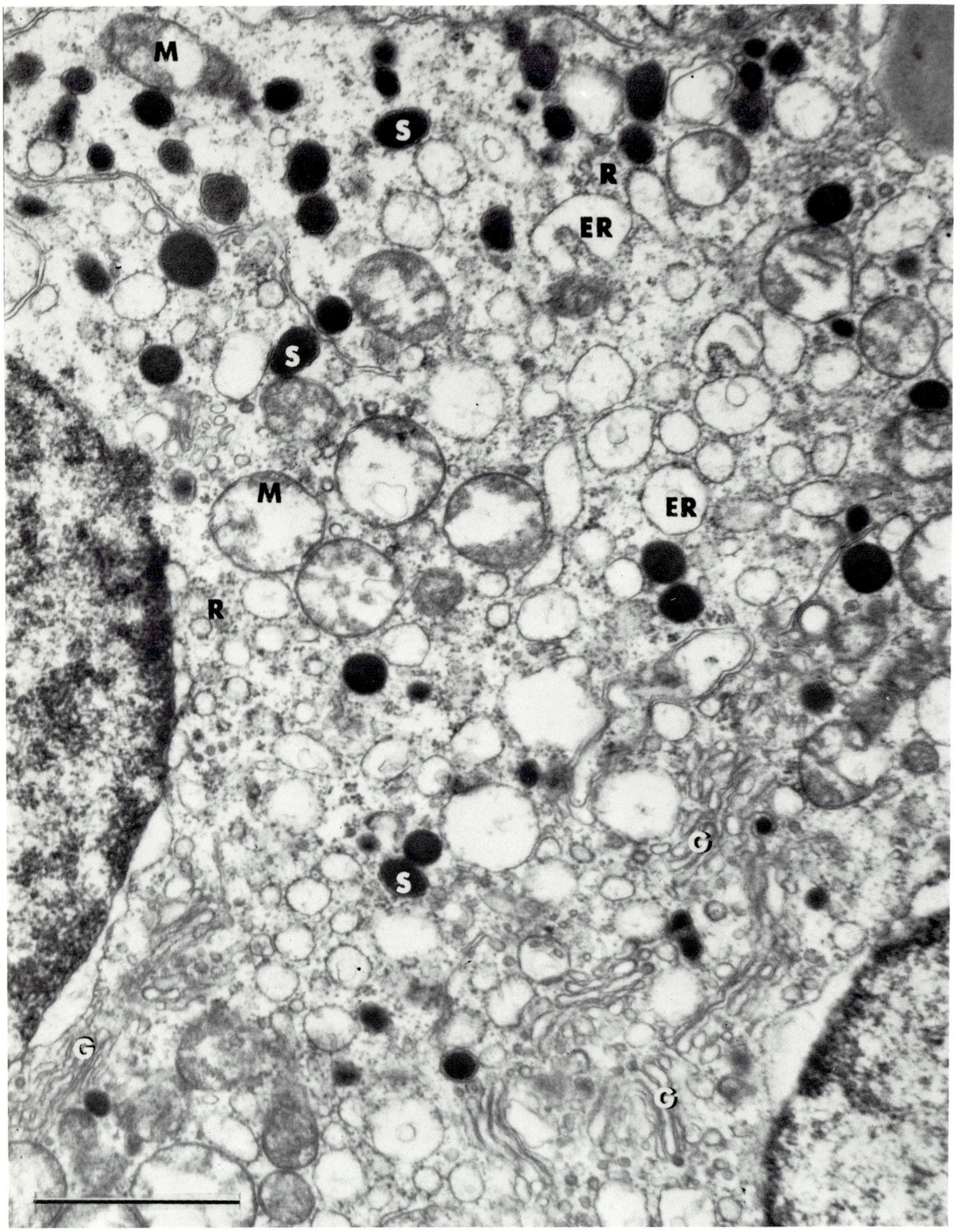

Fig. 10.6. Part of a hypertrophied type I cell of the pars intermedia after renal autograft. Round mitochondria (M) with obliterated cristae, swollen vesiculated endoplasmic reticulum (ER), free ribosomes (R), several well-developed Golgi complexes (G), and oblong granules (S) are shown. × 30,000 (Glutaraldehyde–paraformaldehyde)

posed to a photoperiod which induced growth of a winter white coat in the control animals. After 8 months the grafts were removed, the animals molted, and their coat color became white and remained white during the summer. The results suggest that the pituitary grafts synthesize and secrete MSH. Electron microscopy of these grafts showed active melanotropes (Costoff *et al.,* 1970). An active MSH cell from a renal autograft in a rat is shown in Fig. 10.6. Tixier-Vidal and Picart (1967) also reported the presence of MSH cells in explants of the duck hypophysis.

Summary

There is proof that MSH synthesis occurs in the cells of the pars intermedia. It seems probable that this synthesis takes place in the type I cells and apparently the synthesis and release can take place when the pituitary is not in contact with the hypothalamus. There is also evidence that a pro-ACTH or ACTH might also be synthesized in the cells of the pars intermedia after stress. Perhaps the type II cells synthesize ACTH under these conditions.

OTHER HORMONES OF THE PITUITARY GLAND

From time to time other hormones have been postulated to be produced by the pituitary gland. In some cases the effects of these proposed hormones are so general that bioassays have not been developed. The Li (1968) group have isolated and purified a factor that stimulates the release of lipid from adipose tissue, lipotropin. This hormone consists of fifty-eight amino acids, but a heptapeptide of lipotropin is exactly the same as is found in MSH and ACTH. Whether lipotropin is of pars intermedia or pars distalis origin is not known since the pars intermedia lobes were not removed prior to extraction. Zileli *et al.* (1968) have reported the localization of a hypocalcemic factor in the pituitary. The cellular source of these factors have not been identified.

REFERENCES

Atwell, W. J. (1919). On the nature of the pigmentation changes following hypophysectomy in the frog larva. *Science* **49,** 48.

Bailif, R. N. (1938). Microscopic changes in the hypophysis of the albino rat following exposure to cold and their relationship to the physiology of secretion. *Amer. J. Anat.* **62,** 475.

Barnes, B. G. (1961). Electron microscopic studies on the cytology of the mouse anterior pituitary. *Endocrinology* **71,** 453.

Bryant, W. S. (1916). Sensory elements in the human cerebral hypophysis. *Anat. Rec.* **11**, 25.

Carpenter, E. (1962). Motile cilia in the rat pituitary gland. *Amer. Zool.* **2**, 397.

Carpenter, E. (1971). Fine structure of rat pituitary cells. *Anat. Rec.* **169**, 637.

Costoff, A., Rust, C., and Meyer, R. K. (1970). Unpublished results.

Dubois, P., and Girod, C. (1970). Les cellules ciliées de l'antéhypophyse. *Z. Zellforsch. Mikrosk. Anat.* **103**, 502.

Duchen, L. W. (1962). The affects of ingestion of hypertonic saline on the pituitary gland in the rat: A morphological study of the pars intermedia and the posterior lobe. *J. Endocrinol.* **25**, 161.

Ferrer, J. (1956). Histophysiology of pituitary cleft and colloid cysts in adenohypophysis of rat. Changes after gonadectomy and adrenalectomy. *J. Endocrinol.* **13**, 349.

Gordon, M., ed. (1953). "Pigment Cell Growth." Academic Press, New York. p. 58.

Gosbee, J. L., Kraicer, J., Kastin and Schally, A. V. (1970). Functional relationship between the pars intermedia and ACTH secretion in the rat. *Endocrinology* **86**, 560.

Howe, A., and Maxwell, D. S. (1968). Electron microscopy of the pars intermedia of the pituitary gland in the rat. *Gen. Comp. Endocrinol.* **11**, 169.

Itano, M. (1936). Über den feineren Bau der Pars Intermedia der Hypophysis Cerebri der weissen Ratten. *Acta Anat. Nippon* **9**, 950.

Iturriza, F. C., and Koch, O. R. (1964). Histochemical localization of some α-melanocyte stimulating hormone amino acids in the pars intermedia of the toad pituitary. *J. Histochem. Cytochem.* **12**, 45.

Kastin, A. J., and Schally, A. V. (1966). MSH activity in pituitaries of rats treated with hypothalamic extracts. *Gen. Comp. Endocrinol.* **7**, 452.

Kastin, A. J., and Schally, A. V. (1967). Autoregulation of release of melanocyte stimulating hormone from the rat pituitary. *Nature (London)* **213**, 1238.

Kastin, A. J., Schally, A. V., Viosca, S., and Miller, M. C. (1969). MSH activity in plasma and pituitaries of rats after various treatments. *Endocrinology* **84**, 20.

Kobayashi, Y. (1964). Functional morphology of the pars intermedia of the rat hypophysis as revealed by electron microscopy. 1. Ultrastructural changes after dehydration, *Gunna Symp. Endocrinol.* **1**, 173.

Kobayashi, Y. (1965). Functional morphology of the pars intermedia of the rat hypophysis as revealed by electron microscopy. II. Correlation of the pars intermedia with the hypophysio-adrenal Axis. *Z. Zellforsch. Mikrosk. Anat.* **68**, 155.

Kraicer, J., Bencosme, S. A., and Gosbee, J. L. (1971). The pars intermedia and ACTH secretion in the rat. *Fed. Proc., Fed. Amer. Soc. Exp. Biol.* **30**, 533.

Kurosumi, K., Matsuzawa, T., and Shibasari, S. (1961). Electron microscopic studies on the fine structure of the pars nervosa and pars intermedia and their morphological interrelationships in the normal rat hypophysis. *Gen. Comp. Endocrinol.* **1**, 433.

Kurosumi, K., Matsuzawa, T., and Fujie, E. (1962). Histological and histochemical studies on the rat pituitary pars intermedia. *Arch. Histol.* **22**, 209.

Legait, H. (1964). "Recherches Histophysiologique sui le Lobe Intermediare de l'Hypophyse." Soc. d'Impressions Typographique, Nancy.

Lewis, D., Lee, F. D., and Astwood, E. B. (1937). Some observations on intermedia. *Johns Hopkins Hosp. Bull.* **61**, 198.

Li, C. H. (1968). Current concepts on the chemical biology of pituitary hormones. *Perspect. Biol. Med.* Vol. **11**, 498.

Millhouse, E. W., Jr. (1967). Additional evidence of ciliated cells in the adenohypophysis. *J. Microsc. (Paris)* **6**, 671.

Paiz, C. (1970). Ultrastructure of the human pituitary cleft follicle. *Fed. Proc., Fed. Amer. Soc. Exp. Biol.* **29**, 362 (abstr.).

Phifer, R. F., and Spicer, S. S. (1970). Specific immunologic demonstration of ACTH in the pars intermedia of the adenohypophysis. *Fed. Proc., Fed. Amer. Soc. Exp. Biol.* **29**, 509 (abstr.).

Porte, A., Klein, M. J., Stoeckel, M. E., and Stutinsky, F. (1971). Sur l'existence de cellules de type "Corticotrope" dans la pars intermedia de l'hypophyse du rat. *Z. Zellforsch. Mikrosk. Anat.* **115**, 60.

Purves, H. D. (1961). Morphology of the hypophysis related to its functions: *In* "Sex and Internal Secretions" (W. Young, ed.), 3rd ed., Vol. 1, p. 161. Williams & Wilkins, Baltimore, Maryland.

Raftery, A. T. (1969). Enzyme histochemistry of the pars intermedia of ox pituitary gland. *J. Anat.* **105**, 307.

Ramón y, Cajal, S. (1894). Alcunas contribuciones al conoscimento de los ganglios de cerebro III hipofisis. *Ann. Soc. Exp. Hist. Natur.* **2**, 3.

Rasmussen, A. T. (1929). The percentage of the different types of cells in the male adult human hypophysis. *Amer. J. Pathol.* **5**, 263.

Romeis, B. (1940). Inkretorische organ. *In* "Handbuch der mikroskopischen Anatomie des Menschen" (F. von Möllendorff, ed.), Vol. 7, Part 4, p. 68. Springer-Verlag, Berlin and New York.

Rust, C. C., and Meyer, R. K. (1968). Effect of pituitary autografts on hair color in short-tailed weasel. *Gen. Comp. Endocrinol.* **11**, 548.

Taleisnik, S., Orias, R., and de Olmos, J. (1966). Topographic distribution of melanocyte stimulating hormone releasing factor in rat hypothalamus. *Proc. Soc. Exp. Biol. Med.* **122**, 325.

Tixier-Vidal, A., and Picart, R. (1967). Etude quantitative par radioautographic au microscope électronique d'utilisation de la DL leucine ^{3}H par les cellules de l'hypophyse du canard an culture organotypiqe. *J. Cell Biol.* **35**, 50.

Trautman, A. (1911). Hypophysis cerebri. *In* "Handbuch der vergleichenden-mikroskopischen Anatomie der Haustiere" (W. Ellenberger, ed.), Vol. 2, p. 149. Parey, Berlin.

Vanha-Perttula, T. (1966). Esterases of the rat adenohypophysis. *Acta Physiol. Scand* **69**, Suppl. 283 108 pp.

Vanha-Perttula, T., and Arstila, A. V. (1970). On the Epithelium of the rat pituitary residual lumen. *Z. Zellforsch. Mikrosk. Anat.* **108**, 487.

Wheatley, D. N. (1967). Cells with two cilia in the rat adenohypophysis. *J. Anat.* **101**, 479.

Wingstrand, K. G. ((1966). Microscopic anatomy, nerve supply and blood supply of the pars intermedia. *In* "The Pituitary Gland" (W. Harris and B. Donovan, eds.), Vol. 3, p. 58. Butterworth, London.

Ziegler, B. (1963). Licht und Elektronenmikroskopische Untersuchungen an Pars Intermedia and Neurohypophyse der Ratte. *Z. Zellforsch. Mikrosk. Anat.* **59**, 486.

Zileli, M. S., Cagler, S., Urunay, G., Guner, T., Muftiroplu, E., and Kanna, G. (1968). Localization of hypocalcemic factor in the pituitary gland. *Experientia* **24**, 1263.

Chapter 11

Hormonal Activities and Biochemical Properties of Isolated Granule Fractions

A gradient centrifugation method for the isolation of secretory granules from rat anterior pituitary glands was described in Chap. 2 (Experimental Procedures). In Chaps. 3–7 the pituitary cell types of the pars distalis are discussed. In the first part of this chapter the biological properties of granules will be described. The relative potencies and percentages of the different hormones recovered in individual granule fractions are given in Tables 11.I and 11.II, respectively. The biochemical properties of the granule fractions are described in the second part of this Chapter.

BIOLOGICAL ACTIVITIES OF GRANULE FRACTIONS

Thyrotropic Hormone Activity. Almost 82% of the TSH activity was recovered in the D1HSP pellet (Tables 11.I and 11.II). These results are not in accord with those of Kamat *et al.* (1960) who fractionated TSH tumors from mice and showed TSH activity was associated with a microsomal fraction from their gradient. Hymer and McShan (1963) observed TSH activity in fractions consisting of small granules and microsomal fragments. Since the thyrotrope is a basophil and basophilic granules are less stable than acidophilic ones, it was expected that more than 4% of the TSH activity would be present in the SA soluble fraction. The reason for this small amount of TSH in the soluble fraction appears to be that TSH granules have more acid mucopolysaccharides and behave

TABLE 11.I

RELATIVE POTENCIES OF THE HORMONIES IN THE ANTERIOR PITUITARY FRACTIONS[a]

Fractions	TSH	ACTH	FSH	LH	STH	LTH
FS1	9.31[b] (5.82–14.85)[c]	42.50 (39.10–46.62)	35.90 (29.21–39.52)	2.08 (1.13–3.84)	69.00 (44.10–108.64)	12.20 (9.70–15.00)
NP	0.00	0.50 (0.40–0.72)	0.36 (0.29–0.40)	0.01	0.28 (0.21–0.43)	0.00
SA	0.37 (0.18–0.76)	8.10 (7.38–9.15)	4.84 (4.03–5.29)	0.48 (0.24–0.82)	13.33 (8.16–21.62)	1.05 (0.85–1.28)
BP + CP	0.02	0.00	0.00	0.00	0.00	0.00
D1HSP	7.62 (4.75–12.10)	22.70 (21.30–24.68)	1.11 (0.71–1.43)	0.02	0.00	0.00
FD2LSP	0.00	1.00 (0.80–1.32)	3.74 (2.82–4.31)	0.16 (0.07–0.38)	4.14 (3.02–5.27)	0.00
FD2HSP	0.14 (0.07–0.21)	4.60 (4.10–5.83)	20.35 (15.85–22.13)	1.07 (0.59–1.91)	0.00	0.00
FE1LSP	0.00	0.00	0.00	0.01	45.92 (28.34–77.57)	0.12 (0.08–0.15)
E2LSP	0.00	0.00	0.00	0.00	1.68 (1.12–2.58)	10.90 (8.92–13.41)
Indices of precision (ranges)	0.34–0.36	0.08–0.10	0.08–0.11	0.19–0.25	0.30–0.35	0.10–0.15
No. of Assays	3	4	4	3	2	2

[a] Results given in milliunits per milligram equivalent of fresh tissue. Relative potencies given as follows:
TSH—adjusted to NIH-S1 Standard, expressed as USP mu = milliunits/mg eq FT
ACTH—adjusted to Third International Standard, mu = milliunits/mg eq FT
FSH, LH, STH and LTH—adjusted to NIH-S1 Standards, mu = milliunits/mg eq FT

[b] Mean relative potencies calculated as arithmetic means.

[c] The 95% confidence limits of the potency determinations.

TABLE 11. II

PERCENTAGES OF HORMONAL ACTIVITIES RECOVERED FROM THE GRADIENT FRACTIONS

Fractions	Percentages[a]					
	TSH	ACTH	FSH	LH	STH	LTH
NP	0.0	1.0	0.0	0.5	0.0	0.0
SA	4.0	19.1	13.5	23.1	19.3	8.6
BP + CP	0.2	0.0	0.0	0.0	0.0	0.0
D1HSP	*82.0*[b]	*53.4*	3.1	1.5	0.0	0.0
FD2LSP	0.0	2.4	10.4	7.7	6.0	0.0
FD2HSP	1.5	10.8	*55.7*	*51.4*	0.0	0.0
FE1LSP	0.0	0.0	0.0	0.0	*66.6*	1.0
E2LSP	0.0	0.0	0.0	0.0	2.4	*89.3*
Total recoveries	87.7	86.7	82.7	84.2	94.3	98.9

[a] Based on the amounts of the FS1 fraction added to the gradient, taken as 100% as given in Table 11. I, line 1.

[b] The maximum percentage recoveries for each of the hormones are in italic type.

more like acidophilic granules (Herlant, 1960). Thus in the current separation, TSH was largely associated with one zone from the gradient which is free of contamination from other major organelles.

Adrenocroticotropic Hormone Activity. The ACTH assay showed that 53% of this hormone was also present with the TSH in the small D1HSP granule pellet. Essentially 22% of the ACTH was found in the SA soluble zone and the FD2HSP fraction. These ACTH granules are somewhat larger than thyrotropic granules. This may account for the minimal amount of ACTH associated with the FD2HSP pellet which consisted mainly of larger size granules. The granule measurements and assay data parallel the results reported by Perdue and McShan (1962). They used a discontinuous gradient and found 40% of the ACTH activity in a small fraction consisting of mixed granules 140–200 mμ in diameter.

Gonadotropic Hormones. Approximately 55% and 51%, respectively, of FSH and LH activities were found in the FD2HSP pellet. Thirteen percent and 23%, respectively, of the FSH and LH activities were present in the SA soluble zone. McShan and Hartley (1965) reported that basophilic granules are not as stable as those from acidophils. This may account for the relatively high recoveries of the gonadotropic hormones in the SA soluble fraction. FSH and LH activities were not found in the FE1LSP or E2LSP large granule fractions. This supports the concept that FSH is not associated with the large, slightly dense bodies found in FSH cells. Since FSH and LH are associated with granules of essentially the same diameters and densities, modified methods will be required for their further separation.

Somatotropic Hormone. The assays of the pituitary fractions demonstrated that 67% of the growth hormone was associated with the FE1LSP granule pellet. The data agree with the assay results of Hymer and McShan (1963), except that more STH activity was recovered in a single zone (FE1LSP) and it was not found in the small granule fractions. In the present work a major part of the granular somatotropin was associated with one gradient zone and pellet.

Mammotropic Hormone. Almost 90% of the prolactin activity was associated with the E2LSP granules which were recovered from the lowest zone of the gradient. This is in contrast to the 25% LTH activity recovered by Hymer and McShan (1963).

Discussion

The assays of the hormonal activities in the SA soluble zone of the gradient gave a shallow slope in the dose–response curve as compared to the steep responses of the granule fractions. It is suggested that this kind of response may have been due to the presence of substances in the SA zone which affect the target response (McShan *et al.,* 1953).

This granule fractionation technique can be used to separate granules from pituitaries of larger animals so that more extensive biological and biochemical studies of the hormone can be made. Granules can be extracted and the hormones they contain purified by electrophoresis or chromatographic techniques. Hodges and McShan (1970) extracted similar granule preparations and separated the hormones by polyacrylamide gel electrophoresis. Other workers have used similar fractionation techniques to obtain granule preparations from bovine pituitary glands. Tesar *et al.* (1969) reported separate preparations of acidophilic and basophilic granules from bovine glands. Krass *et al.* (1969) extracted TSH granules and chromatographed the preparation to obtain a purified bovine TSH. Beef growth hormone and proclactin granules were prepared on a discontinuous density gradient by Labrie *et al.* (1971) and LaBella *et al.* (1971). The latter workers solubilized and chromatographed the preparations on DEAE-cellulose.

BIOCHEMICAL AND PHYSICAL PROPERTIES OF GRANULES

Biochemical Studies of Granules

Biochemical studies on the secretory granules of the anterior pituitary were made by Perdue and McShan (1962), Hymer and McShan (1963), and McShan (1971). These workers conducted biochemical analyses on their large and small granule preparations to determine the biochemical properties and the degree of purity of the isolated granules.

The granule fractions were assayed for enzymes known to be associated with specific cytoplasmic organelles. They were found to be free of succinoxidase and cytochrome oxidase activities. These enzymes are associated with mitochondria (Levey and Roberts, 1957), and their absence in the granule fractions indicates that the preparations were free of mitochondria.

Acid phosphatase was not present in the small or large granule pellets, but was associated with the pellet that contained very small granules and microsomes. It seems likely that this pellet contained Golgi vesicles with which acid phosphatase is known to be associated (Novikoff *et al.*, 1962). Trace amounts of alkaline phosphatase was also shown to be in the microsomal pellet; it was not found in any of the purified granule preparations. This enzyme is usually associated with the endothelium of the capillaries (Melchior and Micuta, 1956). This preparation of microsomes must have consisted in part of microsomes or membranes from endothelial cells.

Glucose-6-phosphatase activity was absent in all preparations except the very small granule fraction of Perdue (1962). This enzyme was found in microsomes (Hers *et al.*, 1951) and indicated slight contamination of the very small granule pellet. Thiamine pyrophosphatase usually associated with the Golgi vesicles (Novikoff and Goldfischer, 1961) was not found in the granule pellets.

Acid protease was not present in these granule fractions, but alkaline protease was found in appreciable quantities in the small basophilic fractions of Perdue and McShan (1962). This suggests that this enzyme may be an integral part of the basophilic granule or it may be absorbed on these small granules. Adams and Smith (1951), Meyer and Clifton (1956), and Ellis (1960) suggested that alkaline protease may be involved in the release or secretion of hormones from small granules. Only alkaline protease was associated with the small basophilic granules, which agrees with the findings of Tesar (1967) who reported 15 times more alkaline protease in the small basophilic granules than in the acidophilic granules from the pituitary glands. Smith and Farquhar (1970) have associated an acid phosphatase with acidophilic granules and suggest that this enzyme may be involved with their release. Labrie *et al.* (1971) reported the presence of an ATPase in their bovine large granule preparation. They suggest that this enzyme may be involved in granule release.

Chemical Studies

The protein content of the large acidophilic granule fraction was 9% of that found in the filtered supernatant (FS1), while the proteins of the basophilic granule fraction was 14% (Hymer, 1962). It is postulated that the hormone accounts for only a small part of the total protein of the secretory granule since protein is also found in the granule membrane. Perhaps the extra protein in the small granule fraction is due to the alkaline protease associated with it. Purves (1961) attributed the PAS-positive response of MSH cells to a glycoprotein carrier protein in the MSH granule. It may be postulated on this basis that the proteins in the granules are hormone (or prohormone), enzymes involved in the release of the granule contents, inert protein, and granule membrane protein.

RNA was not found in the basophilic and acidophilic granules but was present in the microsomal fine granule fraction. This indicates that RNA is not associated with the granules but is present in the microsomal fractions. Added proof that RNA is not associated with granules is provided by Peterson and Weiss (1955), who reported that basophilic granules stain with methylene blue, but when the dye was mixed with ribonuclease the granules still stained.

Acidophilic and basophilic granules in whole tissue were first shown by electron microscopy to be surrounded by membranes (Farquhar and Wellings, 1957). These membranes are 60–80 Å thick and are trilaminar. Perdue (1962) reported that 70% of the organic material of the granule is lipid. LaBella and Brown (1959) reported that 75% of the material in granules is lipid. This compares to 50% lipid in mitochondria and 25% in microsomes. Only saturated fatty acids were found in the granule fractions by Hymer (1962). Phospholipids were found in the small granule fractions and were identified by thin layer chromatography as: phosphatides of choline, serine, ethanolamine, inositol and diacylglycerophosphatides. There were up to 4 times more diacylglycerophosphatides in the basophilic granule fraction than in the whole homogenate (Perdue, 1962). In his granule fractions Tesar (1967) found more phospholipids than cholesterol or glycolipids. Slightly more lecithin, sphingomyelin, and phosphatidyl ethanolamine were present in the acidophilic granules than in the granules from basophils.

TABLE 11. III

DIAMETERS OF SECTIONS OF GRANULES FROM RAT ANTERIOR PITUITARY GLANDS

Diameters from sections of	Cell types[a]					
	TSH	ACTH	FSH	LH	STH	LTH
Intact cells						
Mean	85 ± 9[b]	106 ± 10	133 ± 12	145 ± 13	240 ± 17	318 ± 22
Range	40–145	50–180	75–200	75–235	100–390	250–870
	Fractions from gradient[a,c]					
	D1HSP	D1HSP	FD2HSP	FD2HSP	FE1LSP	E2LSP
Isolated granule pellets						
Mean	89 ± 9[b]	89 ± 9	150 ± 15	150 ± 15	242 ± 16	361 ± 22
Range	35–150	35–150	75–218	75–218	180–380	220–880

[a] Results given in millimicrons.
[b] Standard error of the mean.
[c] Fractions (Figure 1) from which maximum amounts of the hormones were recovered.

Although some chemical studies on the lipid content of the granules have been made, considerably more work is required to clarify the role of lipids in the granules.

Physical Properties of Granules

It was shown by LaBella and Brown (1959) that basophilic granules are relatively unstable. Hartley *et al.* (1960) reported that purified gonadotropic granules were readily soluble in water and physiological saline. In our work only 25% of the gonadotropic activity was found in the soluble fractions. In the experiment by Perdue (1962) diisopropylfluorophosphate (DFP) did not affect the stability of the granules at 4° C, but at 37° C there was a stabilizing effect on the granules. DFP is recognized to inhibit the action of alkaline protease which is known to be present in the basophilic granules (Lewis, 1962). In other experiments it was found that acidophilic granules were rather stable in water whereas basophilic granules are easily solubilized. Ethylenediaminetetraacetic acid (EDTA) caused a disintegration of granules (McShan and Hartley, 1965). Kwa *et al.* (1965) noted that prolactin granules were more stable than STH granules. This fact was borne out in the findings reported in our fractionation studies. On the basis of the above information it is concluded that basophilic granules are less stable than acidophilic ones.

Although ACTH granules were not tested for stability, their outer membrane often appears to be broken in electron micrographs. Gonadotropic granules are also unstable but the basophilic thyrotropic granules, because of their high mucopolysaccharide content, are rather firm. In our studies prolactin granules were shown to be more stable than granules from somatotropes. These solubility and stability studies also indicate the importance of determining the nature and role of lipids in the granules and granule membranes.

Diameters of Secretory Granules

Farquhar and Rinehart (1954a,b) were the first to report the maximum diameters of granules in different cell types of the rat pituitary gland. Our values (Table 11.III) of granule diameters are similar to those reported in their early papers. The maximum diameters of the granules provide the most significant information for the identification of the different pituitary cell types. Although there are species variation in the maximum granule diameters, generally they are not very great.

Granule measurements from sections of cells in whole tissue agreed (Table 11.III) whith those values obtained from sections of granule pellets having most of the given hormone activities. For example, granules in TSH cells were similar in size to those found in the D1HSP pellet. This was the fraction with which a major part of the TSH activity was associated. Granules in TSH cells were as much as 40 mμ greater after PTU treatment and granules from LH cells were about 20 mμ greater in diameter after castration. After α-ethyltryptamine treatment most granules in all cell types were slightly larger but those in ACTH cells were 20 mμ greater in diameter. Whether the Golgi has been directly affected by these treatment is a matter of speculation.

REFERENCES

Adams, E., and Smith, E. L. (1951). Proteolytic activity of pituitary extracts. *J. Biol. Chem.* **191**, 651.

Ellis, S. (1960). Pituitary proteinase. I. Purification and action on growth hormone and prolactin. *J. Biol. Chem.* **235**, 1694.

Farquhar, M. G., and Rinehart, J. F. (1954a). Electron microscopic studies of the anterior pituitary gland of castrate rats. *Endocrinology* **54**, 516.

Farquhar, M. G., and Rinehart, J. F. (1945b). Cytologic alterations in the anterior pituitary gland following thyroidectomy; An electron microscopic study. *Endocrinology* **55**, 857.

Farquhar, M. G., and Wellings, S. R. (1957). Electron microscopic evidence suggesting secretory granule formation within the Golgi apparatus. *J. Biophys. Biochem. Cytol.* **3**, 319.

Hartley, M. W., McShan, W. H., and Ris, H. (1960). Isolation of cytoplasmic pituitary granules with gonadotropic activity. *J. Biophys. Biochem. Cytol.* **7**, 209.

Herlant, M. (1960). Etude critique de deux techniques nouvelles destinées à mettre en évidence les différentes catégories céllulairs présentes dans la glande pituitaire. *Bull. Microsc. Appl.* **10**, 37.

Hers, H. G., Berthet, J., Berthet, L., and de Duve, C. (1951). Le Système hexose phosphatastique. III. Localisation intracellaire des ferments par centrifugation fractionae. *Bull Soc. Chim. Biol.* **33**, 21.

Hodges, D. R., and McShan, W. H. (1970). Electrophoretic separation of hormones associated with secretory granules from rat anterior pituitary glands. *Acta. Endocrinol. (Copenhagen)* **63**, 378.

Hymer, W. C. (1962). The isolation of rat pituitary granules and the study of their biochemical properties and hormonal activities. Ph.D. Thesis, University of Wisconsin, Madison.

Hymer, W. C., and McShan, W. H. (1963). Isolation of rat pituitary granules and study of their biochemical properties and hormonal activities. *J. Cell Biol.* **17**, 76.

Kamat, V. M., Wallach, D. F., Crigler, J. F., and Ladman, A. J. (1960). The intracellular localization of hormone activity in transplantable thyrotropin secreting pituitary tumor in mice. *J. Biophys. Biochem. Cytol.* **7**, 219.

Krass, M. E., LaBella, F. S., and Mailhet, R. (1969). Bovine thyrotropin, molecular weight of the hormone derived from soluble and granular pituitary fractions and preparation of an of an active fragment by enzyme hydrolysis. *Endocrinology* **84**, 1257.

Kwa, H. G., Van der Bent, E. M., Feltkamp, C. A., Rümke, P., and Bloemendal, H. (1965). Studies on hormones from the anterior pituitary gland. I. Identification and isolation of growth hormone and prolactin from the granular fraction of bovine pituitary. *Biochem. Biophys. Acta* **111**, 447.

LaBella, F. S., and Brown, J. H. (1959). Cell fractionation of anterior pituitary glands from beef and pork. *J. Biophys. Biochem. Cytol.* **5**, 17.

LaBella, F. S., Krass, M., Fritz, W., Vivian, S., Shin, S., and Queen, G. (1971). Isolation of cytoplasmic granules containing growth hormone and prolactin from bovine pituitary. *Endocrinology* **89**, 1094.

Labrie, F., Lemaire, S., Poirier, G., Pelletier, G., and Boucher, R. (1971). Adenohypophyseal secretory granules. *J. Biol. Chem.* **246**, 7311.

Levey, H. A., and Roberts, S. (1957). Influence of thyroid function on the metabolism of the anterior pituitary gland. *Amer. J. Physiol.* **189**, 86.

Lewis, U. J. (1962). Enzymatic transformations of growth hormone and prolactin. *J. Biol. Chem.* **237**, 3141.

McShan, W. H. (1971). Secretory granules from anterior pituitary glands, *Mem. Endocrinol.* **19**, 161.

McShan, W. H., and Hartley, M. W. (1965). Production, storage and release of anterior pituitary hormones. *Ergeb. Physiol., Biol. Chem. Exp. Pharmakol.* **56**, 264.

McShan, W. H., Rozich, R., and Meyer, R. K. (1953). Biochemical properties of fractions obtained from rat anterior pituitary glands by differential centrifugation. *Endocrinology* **52**, 215.

Melchior, J., and Micuta, B. S. (1956). Effects of administration of estrogens upon enzymes of rat pituitary gland. *Cancer Res.* **16**, 520.

Meyer, R. K., and Clifton, K. H. (1956). Effects of diethylstilbestrol on the quantity and intracellular distribution of pituitary proteinase activity. *Arch. Biochem. Biophys.* **62**, 198.

Novikoff, A. B., and Goldfischer, S. (1961). Nucleoside phosphatase activity in the Golgi apparatus and its usefulness for cytological studies. *Proc. Nat. Acad. Sci. U.S.* **47**, 802.

Novikoff, A. B., Essner, E., Goldfischer, S., and Hess, M. (1962). Nucleosidephosphatase activities of cytomembranes. *Symp. Int. Soc. Cell Biol.* **1**, 149.

Perdue, J. F. (1962). Isolation of secretory granules from rat pituitary glands and the study of their hormonal and biochemical properties. Ph.D. Thesis, University of Wisconsin, Madison.

Perdue, J. F., and McShan, W. H. (1962). Isolation and biochemical study of secretory granules from rat pituitary glands. *J. Cell Biol.* **15**, 159.

Peterson, R. K., and Weiss, J. (1955) . Staining of the adenohypophysis with acid and basic dyes. *Endocrinology* **57**, 96.

Purves, H. D. (1961). Morphology of the hypophysis related to its function. In "Sex and Internal Secretions," (W. Young, ed.), 3rd ed., Vol. 1, p. 161. Williams & Wilkins, Baltimore, Maryland.

Smith, R. E., and Farquhar, M. G. (1970). Modulation in nucleoside diphosphatase activity of mammotropic cells of the rat adenohypophysis during secretion. *J. Histochem. Cytochem.* **18**, 237.

Tesar, J. T. (1967). Hormone storage granules in beef pituitary-isolation and composition. *Fed. Proc. Fed. Amer. Soc. Exp. Biol.* **26**, 534 (abstr.).

Tesar, J. T., Koenig, H., and Hughes, D. (1969). Hormone storage granules in beef anterior pituitary. I. Isolation, ultrastructure, and some biochemical properties. *J. Cell Biol.* **40**, 225.

Chapter 12

Synthesis, Storage, and Release of Anterior Pituitary Hormones

The mechanisms involved in the formation and release of secretory granule from a cell are thought to be similar in several different organs: endocrine and exocrine pancreas, salivary glands, adrenal medulla, and anterior pituitary. Results from experiments performed on each of these organs have contributed to our current knowledge on the biosynthesis and release of hormones and digestive enzymes.

FORMATION OF SECRETORY GRANULES

As early as 1914, Ramón y Cajal postulated that secretory granules may be formed within the Golgi complex. Farquhar and Wellings (1957) and Farquhar (1961) reported in their ultrastructural studies an increase in Golgi activity and in granules when the pituitary gland is actively secreting. They also described granules in different stages of formation in the Golgi apparatus.

Caro and Palade (1961, 1964) conducted light and electron microscopic autoradiographic studies on exocrine pancreas cells. Five minutes after injection of leucine-^{3} H the label was found in the endosplasmic reticulum; after 20 minutes the label was in the Golgi apparatus and in 1 hour it was incorporated into the granule. Autoradiographic studies coupled with the fractionation of cytoplasmic organelles by Redman *et al.* (1966) showed that the label was first incorporated in the ribosomes of the rough endoplasmic reticulum which was recovered as rough microsomes after fractionation. They also showed that the protein synthesized on the attached ribosomes of the endoplasmic reticulum moves unidirectionally across the endoplasmic reticulum into the cisternae. This and other evidence indicate that synthesis of protein occurs on the attached ribosome, moves to the cisternae of the endoplasmic reticulum, and then to the Golgi complex.

In further experiments on the pancreas, Jamieson and Palade (1967a,b, 1968a,b) coupled fractionation with autoradiography and found the label first in the rough microsomes, then in the smooth microsomes, next in the condensing vesicles of the Golgi complex, and finally in the secretory granules. They also have shown that there is not a steady rate of synthesis of secretory proteins and movement of them into the cisternae of the endoplasmic reticulum. Zeigel and Dalton (1962) observed "blebs" breaking off from the endoplasmic reticulum and these appeared to be moving toward the Golgi region. Jamieson and Palade (1967a,b) and Flickinger (1969) showed that small agranular vesicles bud-off from the endoplasmic reticulum and fuse with Golgi vesicles (condensing vesicles). The transport of protein from ribosomes to endoplasmic reticulum does not require energy, but the transport of the protein filled vesicle toward the Golgi complex and its meeting with the condensing vesicle of the Golgi does require energy. This energy is in the form of energy-rich adenosine triphosphate (ATP) which is generated by oxidative phosphorylation.

The smooth vesicles which contain the protein finally fuse with the condensing vesicles of the Golgi complex. There are granules in different stages of formation in the Golgi complex. It is postulated that the condensing vesicles may fuse several times to eventually form the mature granule. As early as 1938 Kirkland and Severinghaus postulated that the Golgi apparatus acts as a condensation system to form granules. Dalton and Felix (1956, 1961) reported that the Golgi complex might be involved in removing water from the maturing secretory granules. Histochemical and electron microscopic work by Smith and Farquhar (1970) has implicated nucleoside diphosphatase activity with the formation of secretory granules in the Golgi apparatus in cells of the rat pituitary gland.

Perdue (1962) estimated that 60% of pituitary secretory granules is lipid, consisting of largely phospholipid. The question also arises concerning where the lipoproteins are formed. According to the results of Hokin (1968) and Claude (1968), they are formed in the endoplasmic reticulum. Beams and Kessel (1968) established that at least part of the phospholipid is added to the protein molecule in the Golgi complex.

Since thyrotropin, follicle-stimulating, and luteinizing hormones are glycoproteins, and since it is believed that other pituitary hormones may have some carbohydrate moiety within the secretory granule (Purves, 1961), it is important to establish where the carbohydrate moiety is added to the protein. Rambourg *et al.* (1969), Neutra and Leblond (1969), and Beams and Kessel (1968) cite histochemical and autoradiographic evidence that implicates the Golgi as the organelle in which addition of carbohydrate takes place during the formation of the secretory granules. Within 15 minutes after injection, tritiated glucose is incorporated into the Golgi vesicles. The carbohydrate is synthesized at the immature side of the Golgi complex but it accumulates in the mature side (granule forming side) of the Golgi. The chains of linked sugars are firmly attached to the proteins by covalent bonds. Schenkein and Uhr (1970) have shown that although hexoses are added to the protein in the Golgi complex, glucosamine is added to nascent protein as well as completed polypeptide chains first in the rough endoplasmic reticulum and then in the Golgi complex.

Tixier-Vidal and Gourdji (1970) studied synthesis and renewal of proteins in duck pituitary glands in cluture. They have reported that the rate of labeling is slow and 50% of the newly synthesized protein remains localized in the cytoplasm while the remaining protein is found in granules. The turnover rate of granular protein is relatively fast but that of tissue protein is slower. They found that *de novo* synthesis of prolactin was linear over a 2-week period and estimate the renewal time of newly synthesized prolactin to be 24–48 hours. About 3.5% of the prolactin in the cells is renewed in 1 hour. They believe that synthesis and release are two discontinuous phenomena. McCann and Porter (1969) have reported that granules are released within 2–10 minutes after a releasing factor is injected. Warshawsky *et al.* (1963) estimated that in the exocrine pancreas exportable protein (granules) spends 11.7 minutes in the Golgi and 36 minutes in the cell before it is released. The total life of a granule would be 47.7 minutes in contrast to 62.5 hours for the turnover time of structural protein.

De Virgilis *et al.* (1968) and Couch *et al.* (1969) injected hypothalamic extract and growth hormone releasing factor into rats, sacrificed the animals after different periods of time, and fixed the tissue for electron microscopy. The De Virgilis group reported that granules were released in large numbers up to 30 minutes after median eminence extracts were given. After 1 hour the release of granules had subsided. The Couch group noted release of granules from STH cells within 2½ minutes after the releasing factor was administered. In 15 minutes the STH cells were depleted of granules. Perhaps the discrepancy in these findings is due to nonspecific releasers in the median eminence extracts. Müller and Pecile (1966) found that STH granules were beginning to deplete at 15 minutes and STH in the pituitary gland was very low 1 hour after median eminence extracts were administered, but in 5 hours granulation in somatotropes returned to normal.

RELEASE OF PITUITARY SECRETORY GRANULES

The mechanism of release of secretory granules from the rat anterior pituitary cells has been described by Farquhar (1961). This mechanism of granule discharge is termed reverse pinocytosis, exocytosis, or emiocytosis and was first observed in the pancreas by Palade (1959). Schneider *et al.* (1967) have observed this type of secretion in the adrenal medulla. According to the exocytosis theory of release the outer surface of the granule membrane fuses with the internal surface of the cell membrane. An aperture forms at the site of fusion, and the contents of the granule are released into the intercullular space or capillaries. This type of discharge has been seen in some acidophils and basophils. Pelletier *et al.* (1971) showed that exocytosis occurs only at the cell surface adjacent to the basement membrane. In some basophils after the contents of one granule are expelled, another granule may open into the same place and emit its contents (Fig. 12.1). A similar method of granule release has been shown by Farquhar (1971) to occur in acidophils. The granule membranes left after exocytosis are at first incorporated into the cellular membrane, then are broken down to phosphatidyl inositol and choline, recycled through the endoplasmic reticulum and Golgi, and incorporated into new membranes (Hokin, 1968).

ACTH granules are uniformly dense as they leave the Golgi complex but in certain cases their density decreases as they move toward the periphery of the cell. Furthermore, the membrane surrounding these granules often appeared discontinuous or broken. Discontinuous membranes have also been seen surrounding granules from other cell types. These findings suggest that ACTH leaves the granule intracellularly before it reaches the cell membrane. This may account for the fact that many ACTH granules appear as empty vesicles or contain very little material at the time they reach the cell membrane. Since ACTH is continuously required by animals, it may be continually released intracellularly (Fig. 12.2). Siperstein and Miller (1970) have observed granules containing very little stainable material from other cell types, including TSH cells. Membrane discontinuities have also been reported in rabbit pancreas acinar granules by

Shapiro and Lazarus (1967). They found membrane discontinuities even after calcium was added to the fixative. They concluded that these discontinuities associated with secretory granules are not artifacts. Intracytoplasmic release of FSH has been reported by Herlant (1964). This kind of release of secretory material has also been observed by Lever and Fin-

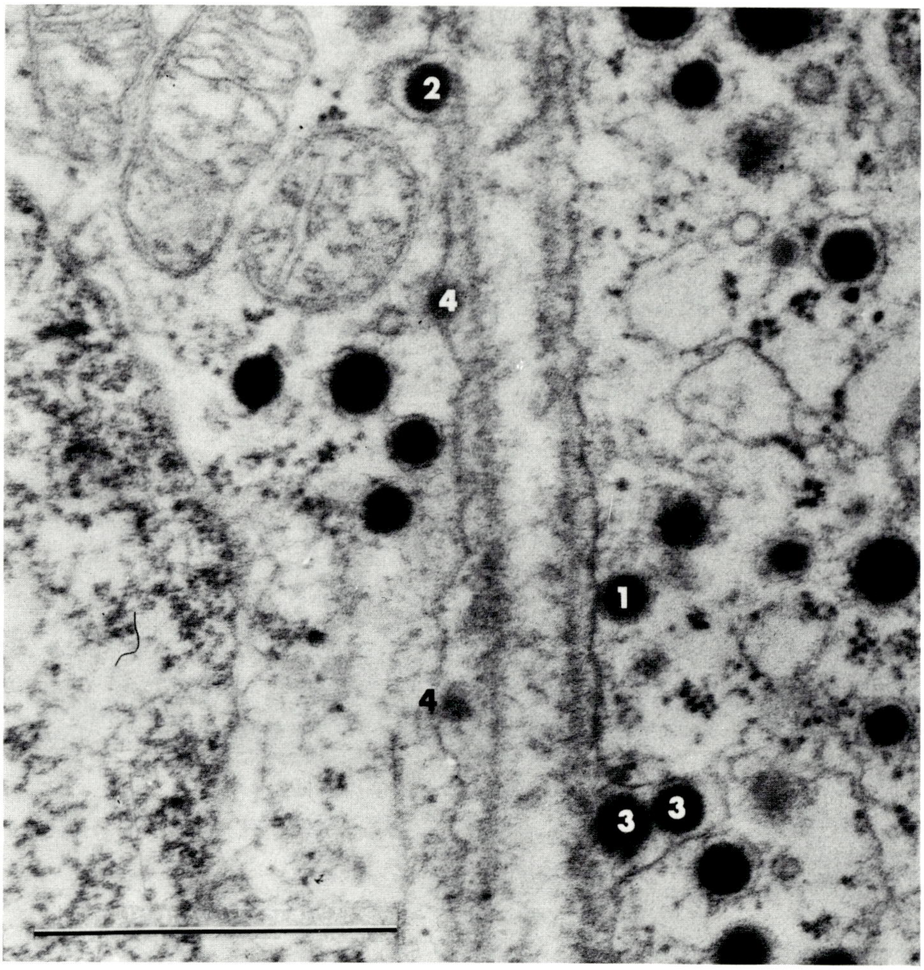

Fig. 12.1. The mechanism of exocytosis is illustrated in this electron micrograph. (1) a granule fuses with the cell membrane, (2) a granule is being released into the intercellular space, (3) two granules are expelled into the same space, (4) contents of released granule are diffusing and are not as apparent as in (2). × 50,000. (Courtesy of Sister Jean Shaefer). (Veronal acetate buffered osmic acid)

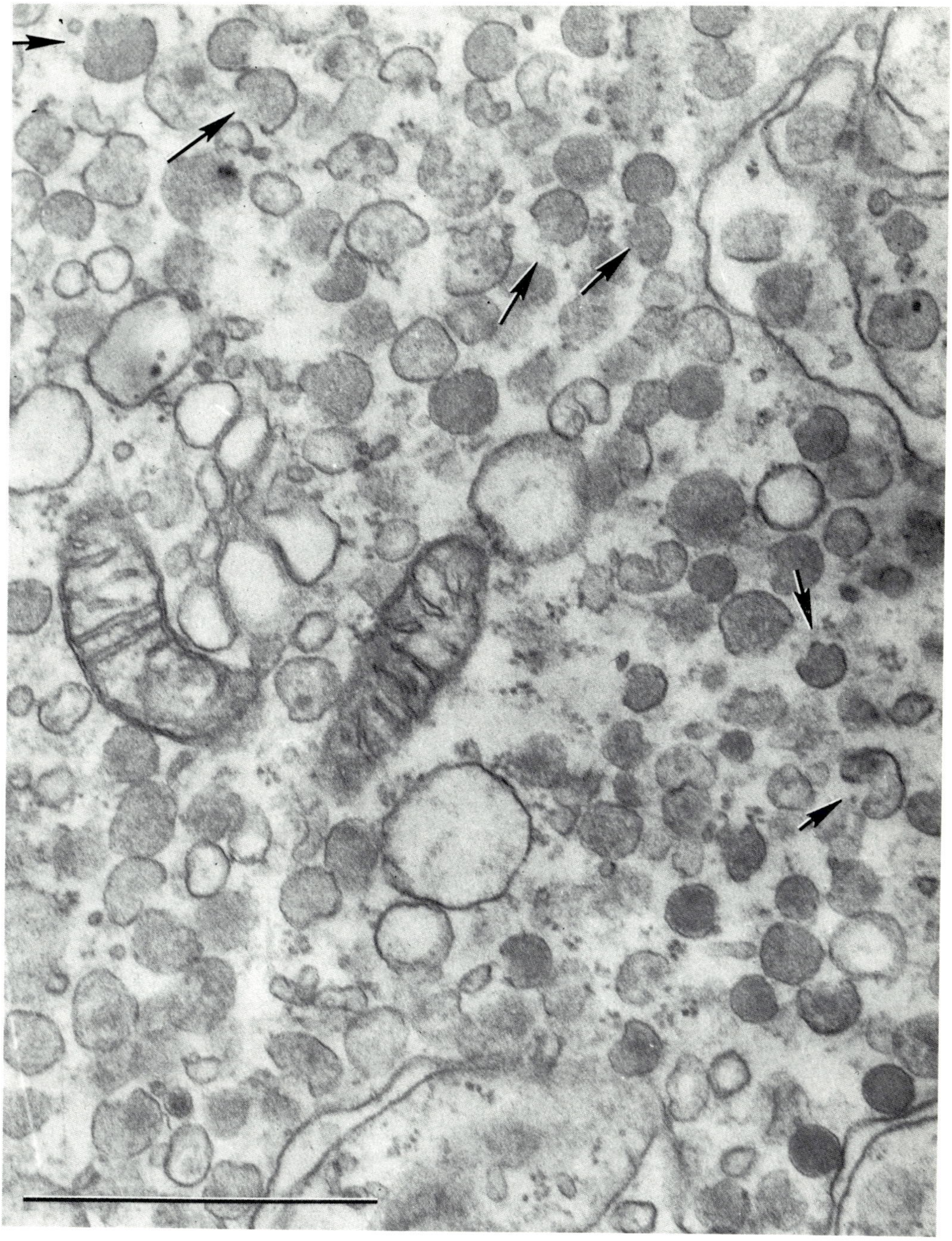

Fig. 12.2. An electron micrograph of a portion of a corticotrope 30 days after adrenalectomy. Many of the granules pictured here are releasing their contents intracellularly (as indicated by arrows). × 52,500. (Veronal acetate buffered osmic acid)

dlay (1966) in the pancreas and the adrenal medulla, and by Hillarp (1959) in the latter.

Although there is evidence that supports the intracytoplasmic release of ACTH and perhaps other hormones there is also a possibility that the uneven density and discontinuous granule membranes may be due to fixation artifacts. Amsterdam and Schramm (1966) reported rapid release of protein from zymogen granules when phosphate buffered osmic acid was used as a fixative. Kurosumi and Kobayashi (1966) also observed these shrunken granules and granules of varying electron density in ACTH cells using Caulfield fixative. Yamada and Yamashita (1967) reported that granules in ACTH cells of pituitaries from mice were better preserved; they used glutaraldehyde fixation. In our studies glutaraldehyde fixation preserved granule membranes detter, especially those of ACTH, but granules of varying densities were still observed. Siperstein and Miller (1970) also noted that granule membranes were better preserved after glutaraldehyde fixation, but they too found granules of varying electron density in ACTH and TSH cells. The question is whether intracytoplasmic release occurs or is an artifact. If intracytoplasmic release is operative it remains to be determined how the hormone passes through the cell membrane.

Another method of secretion may be a microapocrine type where small blebs of cytoplasm containing granules are pinched off from the cell into the intercellular space. This secretion process has been observed in the rabbit pituitary gland by Salazar and Peterson (1964). Lacy *et al.* (1968) have postulated that microtubules may be involved in granule movements within and outside of the cell in the endocrine pancreas. It has also been postulated by Palay (1959) that after synthesis protein passes directly out of the cell through the endoplasmic reticulum. In conclusion the exocytosis type of secretion of granules is well documented but the other methods are speculative and require further experimental work to be proven valid.

The Alteration of Pituitary Cell Membranes by Administration of Releasing Factors

The release of a tropic hormone is governed by a specific releasing factor of hypothalamic origin (Fig. 12.3), and perhaps by a neurohumoral factor, and by a feedback product. These factors act to control synthesis and release of the tropic hormone by acting directly or indirectly on cells of the pars distalis. For a review of releasing factors, consult McCann and Porter (1969).

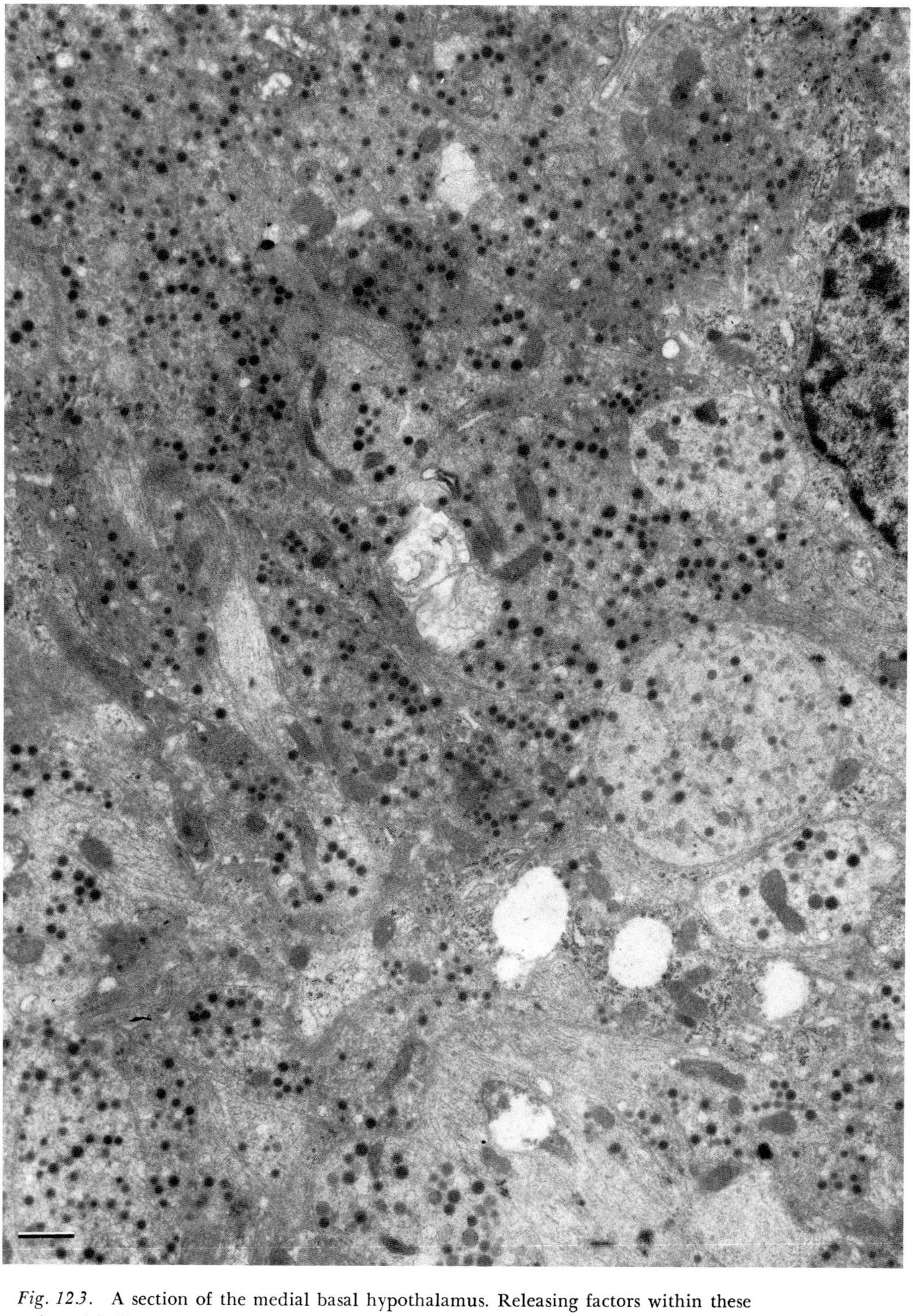

Fig. 12.3. A section of the medial basal hypothalamus. Releasing factors within these axonic end bulbs are secreted into portal vessels of the hypophysis and stimulate release of tropic hormones of the anterior pituitary gland. × 8250. (Glutaraldehyde-paraformaldehyde)

Releasing factors, as the name implies, stimulate release of pituitary hormones from specific pituitary cell types. For instance, corticotropin releasing factor (CRF) stimulates ACTH release from corticotropes and does not act on other types of cells. Other releasing factors such as that for thyrotropin (TRF), and for growth hormone (GRF) will not release ACTH. The releasing factors have been separated and several have been purified; the structure for thyrotropin releasing factor was first elucidated by Schally *et al.* in 1970. Neural factors such as dopamine may stimulate release of the releasing factor which in turn acts on pituitary cells and stimulates release of a tropic hormone. Schneider and McCann (1970) have reported that dopamine stimulates release of luteinizing hormone releasing factor (LHRF) from the median eminence.

These releasing factors may also directly influence synthesis of pituitary hormones either by a primary action or by an action which is secondary to the release of the stored hormone (Kobayashi *et al.,* 1967; Deighton and Meyer, 1969; Coates *et al.* (1971). TRF has been implicated as stimulating synthesis and release of the tropic hormone (Burgus and Guillemin, 1970). Whether a releasing factor stimulates release or synthesis or both is not clearly understood at the present time.

Cyclic adenosine monophosphate (cyclic AMP), "the universal secondary messenger," is thought to be operational in the release mechanism. This may involve a high level of releasing factor (RF) which binds to the pituitary cell membrane and activates adenyl cyclase in the plasma membrane. This enzyme catalyzes the breakdown of ATP to cyclic AMP and inorganic phosphate. The cyclic AMP thus formed initiates exocytosis in a given pituitary cell (Robison *et al.,* 1968). Cyclic AMP also serves to activate a phosphorylase system which breaks down glycogen to glucose 1-phosphate. It has been shown by Migliorini and Antunes-Rodrigues (1970) that releasing factors increase energy metabolism in a given cell by 20%. In this connection stimulation of glycogen breakdown by way of the phosphorylase system is an important action of cyclic AMP. Justisz and de la Llosa (1970), and Zimmerman and Fleischer (1970) have shown that cyclic AMP needs calcium to be operational.

The Effect of Changes in Concentration of Ions on the Cell Membrane

In many cells of the body the intercellular fluid bathing the cells is composed of 155 m*M* sodium and 5 m*M* potassium while the intracellular fluid is made up of 150 m*M* potassium and 20 m*M* sodium. Although the cell membrane is relatively impermeable to sodium, this ion still permeates the cell. A sodium pump is postulated that removes sodium from the cell and at the same time allows potassium to enter. This process requires energy which is supplied by ATP. Adenosine triphosphatase found in the cell membrane catalyzes the breakdown of ATP (Skou, 1965).

Douglas (1967) postulated the existence of the "stimulus-secretion coupling" process of granule release in the adrenal medulla and neuhypophysis. In this theory the release of the granule is initiated by the depolarization of the cell membrane which permits the increased permeability of calcium into the cell.

Samli and Geschwind (1968) reported *in vitro* experiments showing that increased potassium resulted in a greater release of LH. If potassium is increased and luteinizing hormone releasing factor is added to the culture the effect is additive and more LH is released into the medium. It appears that calcium is needed in the medium for the potassium and RF to be operational. These workers also found that calcium or potassium is not a requirement for LH synthesis and that although inhibitors of protein synthesis block LH synthesis they do not effect LH release. Wakabayashi *et al.* (1969) observed similar results but noted that a reduction or omission of sodium increased FSH release. Kraicer *et al.* (1969) reported similar finding for ACTH release but also found that if corticosterone is added to pituitary cultures, ACTH release is decreased. In this case corticosterone may act directly on the cell membrane to inhibit release of ACTH. However, Milligan and Kraicer (1970) showed there was no increase in calcium incorporation into the stimulated pituitary cells.

The *in vitro* experiments of Geschwind (1969, 1971) showed that when potassium levels were increased, cyclic AMP did not increase but high levels of potassium stimulated an increase in hormone release. He also showed that when calcium is absent release of tropic hormone is decreased eventhough adenyl cyclase and cyclic AMP levels were high. Therefore, the latter two compounds were present in high concentrations but hormone release was blocked because of the lack of calcium. It would seem that calcium and not cyclic AMP is the important factor in the release of pituitary tropic hormones.

In the light of the above, it may be postulated that the releasing factor acts by depolarizing the cell membrane, which alters the permeability and allows calcium to enter the cell. This may initiate release of the hormone contained in the granule. To confirm this postulation that there actually is a depolarization of the membrane, it would seem advantageous to check this in pituitary cells. McCann and Porter (1969) reported recordings with extracellular microelectrodes in the anterior pituitary that showed a 1–2 minute depolarization which began 30 seconds after the intravenous injection of hypothalamic extracts to anesthetized rats.

SUMMARY OF SYNTHESIS AND RELEASE OF ANTERIOR PITUITARY HORMONES

A scheme explaining the synthesis and release of pituitary hormones is given in Fig. 12.4.

The information encoded in the DNA of the pituitary cell leaves the nucleus in the form of messenger RNA (mRNA). The releasing factor directly or indirectly increases the amount of mRNA produced. Ribosomes of the endoplasmic reticulum (Fig. 12.5) attach temporarily to a strand of mRNA. A series of ribosomes read the information needed to synthesize a polypeptide chain. Amino acids in the cytoplasm are delivered to the polysome by transfer RNA (tRNA). The protein moves vectorially across the endoplasmic reticulum membrane into the cisternae, and down this tubule to a region where blebs of agranular ER containing the protein and perhaps phospholipid are budded off. The agranular vesicles thus formed are in very close position to the Golgi apparatus (Fig. 12.6) and fuse with small condensing vesicles of the Golgi complex.

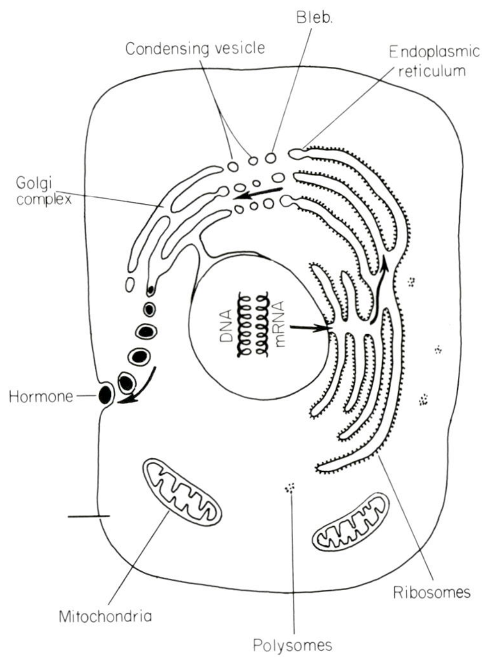

Fig. 12.4. A schematic representation of synthesis, storage, and release of pituitary hormones.

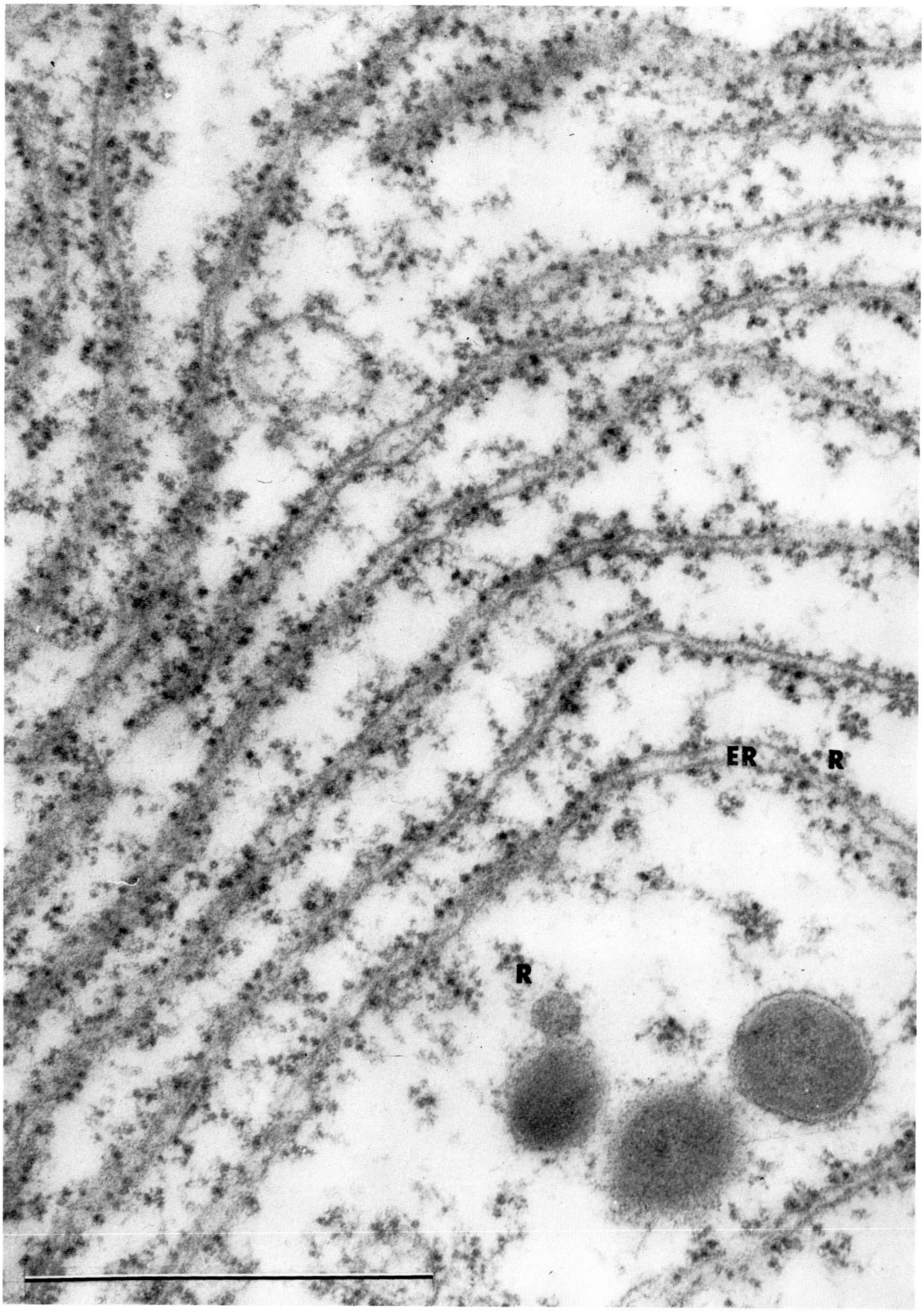

Fig. 12.5. The synthesis of protein begins at the attached ribosomes (R) of the endoplasmic reticulum (ER). × 62,500. (Veronal acetate buffered osmic acid)

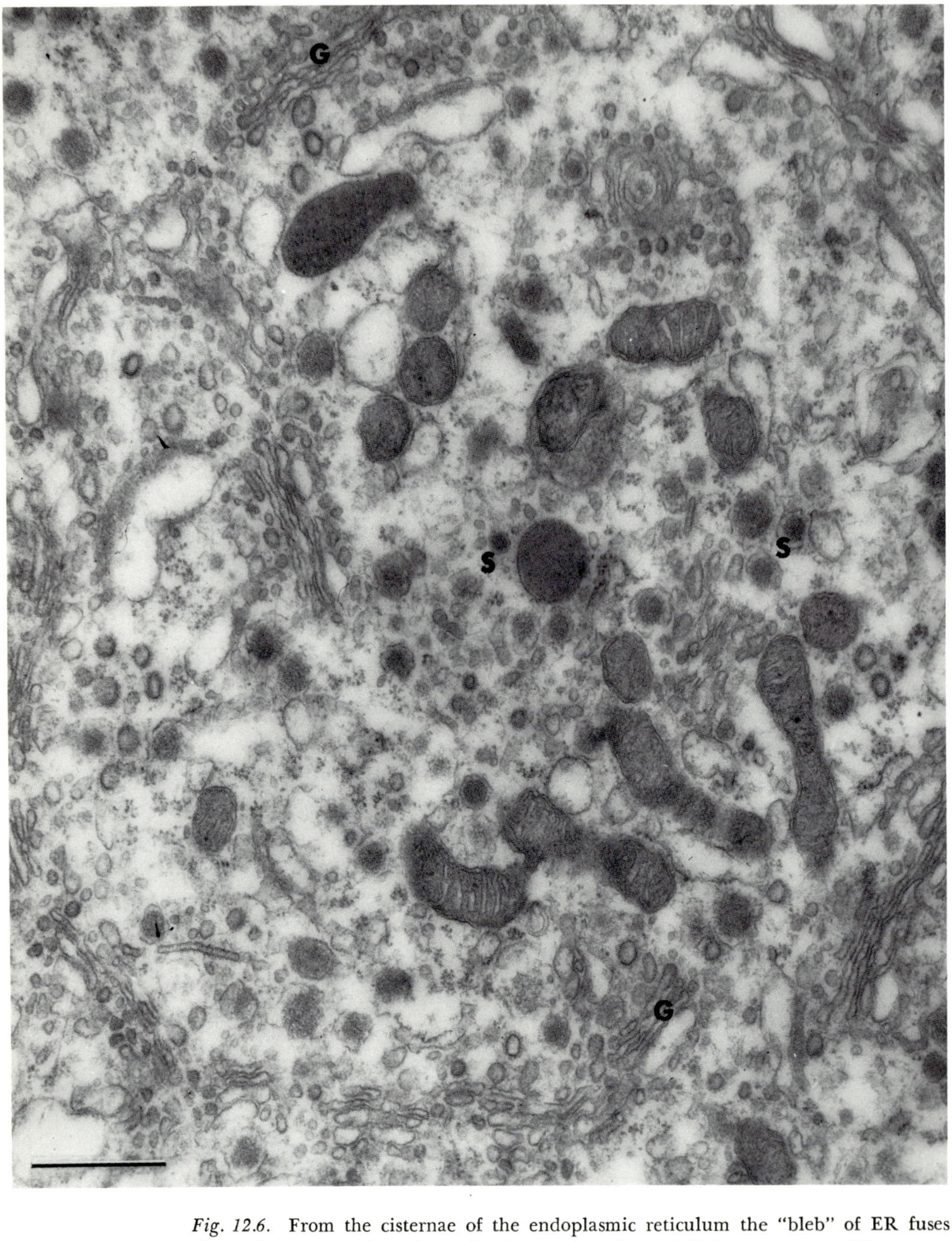

Fig. 12.6. From the cisternae of the endoplasmic reticulum the "bleb" of ER fuses with condensing vesicles of the Golgi complex. Several Golgi complexes (G) are seen around the periphery of the micrograph. Most of these Golgi complexes consist of flattened saccules and small vesicles. Secretory granules (S) in various stages of formation are seen in the center of the micrograph. $\times$ 20,000. (Veronal acetate buffered osmic acid)

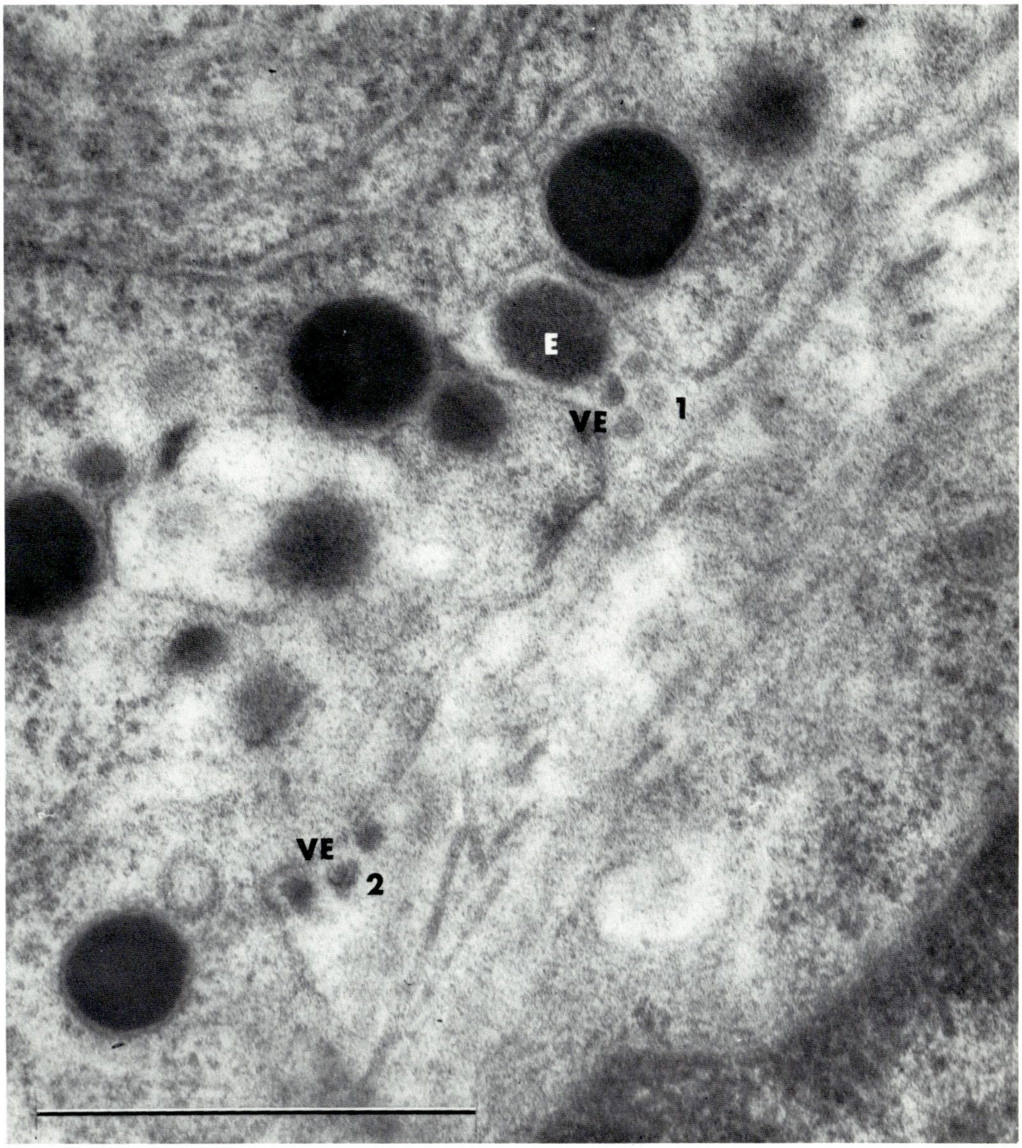

Fig. 12.7. Exocytosis (E) is shown occurring at the cell membrane of a somatotrope. Note the small vesicles (VE) (whose origin may be the median eminence) in the intercellular space near the extruded granule (1) and near the cell membrane (2). × 60,000. (Glutaraldehyde-paraformaldehyde)

In the Golgi complex the molecules of hormone are condensed, water is removed and, if the hormone is destined to be a glycoprotein, carbohydrate and perhaps more lipid are added to the hormone and granule, respectively. The releasing factor or a secondary factor causes mature granules to move away from the Golgi complex. The releasing factor (RF) binds to the cell membrane causing a depolarization of the cell membrane and which results in an increase in potassium entering the cell. Calcium moves into the cell and exocytosis occurs (Fig. 12.7). The RF acting on the cell membrane also causes release of adenyl cyclase from the cell membrane which catalyzes the breakdown of ATP to form cyclic AMP. The cyclic AMP activates the phosphorylase system and thereby enhances the breakdown of glycogen and increases carbohydrate metabolism. Intermediary metabolism may proceed via the triose phosphate pathway but more probably takes place by the hexose monophosphate shunt (Packman and Robins, 1970). This pathway not only provides the reduced nicotinamide adenine dinucleotide phosphate (NADPH) required for phospholipogenesis but also is essential in providing pentoses necessary for increased RNA, especially mRNA, and DNA synthesis.

REFERENCES

Amsterdam, A., and Schramm, M. (1966). Rapid release of the zymogen granule protein by osmium tetroxide and its retention during fixation by glutaraldehyde. *J. Cell Biol.* **29**, 199.

Beams, H. W., and Kessel, R. G. (1968). The Golgi apparatus: Structure and function. *Int. Rev. Cytol.* **23**, 241.

Burgus, R., and Guillemin, R. (1970). Hypothalamic thyrotropin releasing factor. *In* "Hypophysiotropic Hormones of the Hypothalamus" (J. Meites, ed.), p. 227 Williams & Wilkins, Baltimore, Maryland.

Caro, L. G. (1961). Electron microscopic radioautography of thin sections: The Golgi zone as a site of protein concentration in pancreatic acinar cells. *J. Biophys. Biochem. Cytol.* **10**, 37.

Caro, L. G., and Palade, G. E. (1964). Protein synthesis, storage and discharge in the pancreatic exocrine cell. *J. Cell Biol.* **20**, 473.

Claude, A. (1968). Interrelation of cytoplasmic membranes in mammalian liver cells: Endoplasmic reticulum and Golgi complex. *J. Cell Biol.* **39**, 25a.

Coates, P. W., Ashby, E. A., Krulich, L., Dhariwal, A. P., and McCann, S. M. (1971). Fine structure of somatotrophs in the adenohypophysis incubated with growth hormone releasing factor. *Anat. Rec.* **169**, 298.

Couch, E. F., Arimura, A., Schally, A. V., Saito, M., and Sawano, S. (1969). Electron microscopic studies of somatotrophs of rat pituitary after injection of purified growth hormone releasing factor. *Endocrinology* **85**, 1084.

Dalton, A. J. (1961). Golgi apparatus and secretion granules. *In* "The Cell" (J. Brachet and A. E. Mirsky, eds.), Vol. 2, p. 603. Academic Press, New York.

Dalton, A. J., and Felix, M. D. (1956). A comparative study of the Golgi complex. *J. Biophys. Biochem. Cytol.* **2**, Suppl., p. 79.

Deighton, K. J., and Meyer, R. K. (1969). Synthesis of ovulating hormone in organ culture. *Endocrinology* **84**, 778.

De Virgilis, G., Meldolesi, J., and Clementi, F., (1968). Ultrastructure of growth hormone producing cells of rat pituitary after injection of hypothalamic extracts. *Endocrinology* **83**, 1278.

Douglas W, W. (1967). Secretory granule release from the neurohypophysis. *In* "Neurosecretion" (F. Stutinsky, ed.), p. 178. Springer-Verlag, Berlin and New York.

Farquhar, M. G. (1961). Origin and fate of secretory granules in cells of the anterior pituitary gland. *Trans. N. Y. Acad. Sci.* [2] **23**, 346.

Farquhar, M. G. (1971). Processing of secretory products by cells of the anterior pituitary gland. *Mem. Soc. Endocrinol.* **19**, 79.

Farquhar, M. G., and Wellings, S. R. (1957). Electron microscopic evidence suggesting secretory granule formation within the Golgi apparatus. *J. Biophys. Biochem. Cytol.* **3**, 319.

Flickinger, C. J. (1969). The pattern of growth of the Golgi complex during fetal and postnatal development of the rat epididymis. *J. Ultrastruct. Res.* **27**, 344.

Geschwind, I. I. (1969). Mechanism of action of releasing factors. *In* "Frontiers in Neuroendocrinology" (W. F. Ganong and L. Martini, eds.), p. 389. Oxford Univ. Press, London and New York.

Geschwind, I. I. (1971). Mechanisms of release of anterior pituitary hormones—Studies *in vitro*. *Mem. Soc. Endocrinol.* **19**, 221.

Herlant, M. (1964). The cells of the adenohypophysis and their functional significance. *Int. Rev. Cytol.* **17**, 299.

Hillarp, N. A. (1959). The construction and functional organization of the autonomic innervation apparatus. *Acta Physiol. Scand., Suppl.* **157**, 46.

Hokin, L. E. (1968). Dynamic aspects during protein synthesis. *Int. Rev. Cytol.* **23**, 187.

Jamieson, J. D., Palade, G. E. (1967a). Intercellular transport of secretory proteins in the pancreatic exocrine cell. I. Role of the peripheral elements of the Golgi complex. *J. Cell Biol.* **34**, 577.

Jamieson, J. D., and Palade, G. E. (1967b). Intercellular transport of secretory proteins in the pancreatic exocrine cell. II. Transport to condensing vacuoles and zymogen granules. *J. Cell Biol.* **34**, 597.

Jamieson, J. D., and Palade, G. E. (1968b). Intercellular transport of secretory proteins in the pancreatic exocrine cell. III. Dissociation of intracellular transport from protein synthesis. *J. Cell Biol.* **39**, 580.

Jamieson, J. D., and Palade, G. E. (1968b). Intercellular transport of secretory proteins in the pancreatic exocrine cell. IV. Metabolic requirement. *J. Cell Biol.* **39**, 589.

Justisz, M., and de la Llosa, M. P. (1970). Requirement of Ca^{++} and Mg^{++} ions for the *in vitro* release of follicle stimulating hormone from rat pituitary glands and its subsequent biosynthesis. *Endocrinology* **86**, 761.

Kirkland, H., and Severinghaus, A. E. (1938). A review of the Golgi apparatus. *Anat. Rec.* **70**, 413.

Kobayashi, T., Kigawa, T., Mizuno, M., Amenori, Y., Watanabe, T., and Ichikawa, T. H. (1967). Follicle stimulating hormone releasing factor and luteinizing hormone releasing factor in rat hypothalamic extract. *Endocrinol. Jap.* **14**, 101.

Kraicer, J., Milligan, J. V., Gosbee, J. L., Conrad, R. G., and Branson, C. M. (1969). Potassium, corticosterone and adrenocorticotropic hormone release. *Science* **164**, 426.

Kurosumi, K., and Kobayashi, Y. (1966). Corticotrophs in the anterior pituitary glands of normal and adrenalectomized rats as revealed by electron microscopy. *Endocrinology* **78**, 745.

Lacy, P. E., Howell, S. L., Young, D. A., and Fink, C. J. (1968). New hypothesis of insulin secretion. *Nature (London)* **219**, 1177.

Lever, J. D., and Findlay, J. A. (1966). Similar structural basis for the storage and release of secretory material in adreno-medullary and pancreatic cells. *Z. Zellforsch. Mikrosk. Anat.* **74**, 317.

McCann, S. M., and Porter, J. C. (1969). Hypothalamic-pituitary stimulating and inhibiting hormones. *Physiol. Rev.* **49**, 240.

Migliorini, R. H., and Antunes-Rodrigues, J. (1970). Effect of hypothalamic crude extract on the utilization of D (^{14}C) -glucose by the anterior pituitary. *Experientia* **26**, 191.

Milligan, J. V., and Kraicer, J. (1970). Effect of various secretagogues on Ca-45 space in the adenohypophysis. *Fed. Proc., Fed. Amer. Soc. Exp. Biol.* **29**, 312 (abstr.).

Müller, E., and Pecile, A. (1966). Growth hormone resynthesis in the pituitary after depletion induced by a hypothalamic extract. *Experientia* **22**, 108.

Neutra, M., and Leblond, C. P. (1969) . The Golgi apparatus. *Sci. Ameri.* **220**, 100.

Packman, P. M., and Robins, E. (1970). Quantitative enzymatic studies in individual pituitary lobes in various endocrine states. *Endocrinology* **87**, 13.

Palade, G. E. (1959). Functional changes in the structure of cell components. In "Subcellular Particles" (T. Hayashi, ed.) , p. 64–83. Ronald Press, New York.

Palay, S. L. (1959). The morphology of secretion. In "Frontiers of Cytology" (S. L. Palay, ed.). pp. 335–36, Yale University Press, New Haven, Conn.

Pelletier, G., Peillon, F., and Vila-Porcile, E. (1971). An ultrastructural study of sites of granule extrusion in the anterior pituitary of the rat. *Z. Zellforsch. Mikrosk. Anat.* **115**, 501.

Perdue, J. F. (1962). Isolation of secretory granules from rat pituitary glands and the study of their hormonal and biochemical properties. Ph.D. Thesis, University of Wisconsin, Madison.

Purves, H. D. (1961). Morphology of the hypothysis related to its function. In "Sex and Internal Secretions" (W. Young, ed.) , 3rd ed., vol. 1, p. 161 Williams & Wilkins, Baltimore, Maryland.

Rambourg, A., Hernandez, W., and Leblond, C. P. (1969). Detection of complex carbohydrates in the Golgi apparatus of rat cells. *J. Cell Biol.* **40**, 395.

Ramón y Cajal, S. (1914). The role of the Golgi apparatus. *Trab. Lab. Invest. Biol. Univ. Madrid* **12**, 127.

Redman, C. M., Siekevitz, P., and Palade, G. E. (1966). Synthesis and transfer of amylase in pigeon pancreatic microsomes. *J. Biol. Chem.* **241**, 1150.

Robison, G. A., Butcher, R. W., and Sutherland, E. W. (1968). Cyclic AMP. *Annu. Rev. Biochem.* **37**, 149.

Salazar, H., and Peterson, R. P. (1964). Morphologic observations concerning the release and transport of secretory products in the adenohypophysis. *Amer. J. Anat.* **115**, 199.

Samli, M. H., and Geschwind, I. I. (1968) . Some effects of energy-transfer inhibitors and of Ca^{++}–free of K^{+}–enhanced on the release of luteinizing hormone (LH) from the rat pituitary gland *in vitro*. *Endocrinology* **82**, 225.

Schally, A. V., Nair, R. M., Barrett, J. F., Bowers, C. Y., and Folkers, F. (1970). The structure of hypothalamic thyrotropin releasing factor. *Fed. Proc., Fed. Amer. Soc. Exp. Biol.* **29**, 470 (abstr.).

Schenkein, I., and Uhr, J. W. (1970). Immunoglobulin synthesis and secretion. I. Biosynthetic studies of the addition of the carbohydrate moieties. *J. Cell Biol.* **46**, 42.

Schneider, F. H., Smith, A. D., and Winkler, H. (1967). Secretion from the adrenal medulla, biochemical evidence for exocytosis. *Brit. J. Pharmacol. Chemother.* **31**, 94.

Schneider, H. P. G., and McCann, S. M. (1970). Release of LH-releasing factor (LRF) into the peripheral circulation of hypophysectomized rats by dopamine and its blockage by estradiol. *Endocrinology* **87**, 249.

Shapiro, S. H., and Lazarus, S. S. (1967). Membrane discontinuities and secretory granule formation in rabbit pancreatic acinar cells. *Exp. Mol. Pathol.* **6**, 320.

Siperstein, E. R., and Miller, K. J. (1970). Further cytophysiologic evidence for the identity of the cells that produce adrenocorticotrophic hormone. *Endocrinology* **86**, 451.

Skou, J. C. (1965). Enzymatic basis for active transport of sodium and potassium across the cell membrane. *Physiol. Rev.* **45**, 596.

Smith, R. E., and Farquhar, M. G. (1970). Modulation in nucleoside diphosphatase activity of mammotropic cells of the rat adenohypophysis during secretion. *J. Histochem. Cytochem.* **18**, 237.

Tixier-Vidal, A., and Gourdji, D. (1970). Synthesis and renewal of proteins in duck anterior hypophysis in organ culture. *J. Cell Biol.* **46**, 130.

Wakabayashi, K., Kamberi, I. A., and McCann, S. M. (1969). *In vitro* response of the rat pituitary to gonadotropin releasing factors and to ions. *Endocrinology* **85**, 1046.

Warshawsky, H., Leblond, C. P., and Droz, B. (1963). Synthesis and migration of proteins in the cells of the the exocrine pancreas as revealed by specific activity determination from radioautography. *J. Cell Biol.* **16**, 2.

Yamada, Y., and Yamashita, K. (1967). An electron microscopic study on the possible site of production of ACTH in the anterior pituitary of mice. *Z. Zellforsch. Mikrosk. Anat.* **80**, 29.

Zeigel, R. F., and Dalton, A. J. (1962). Speculations based on the morphology of the Golgi systems in several types of protein-secreting cells. *J. Cell Biol.* **15**, 45.

Zimmerman, G., and Fleischer, N. (1970). Role of calcium ions in the relase of ACTH from rat pituitary tissue *in vitro*. *Endocrinology* **87**, 426.

Author Index

Numbers in italics refer to the pages on which the complete references are listed.

T

Subject Index